NemoFactory
(10359) 1F, 52-11, Ilsan-ro 441 beon-gil, Ilsandong-gu, Goyang-si, Gyeonggi-do, Korea
T_ 82-31-915-1511 / **F_** 82-31-915-1520 / **E_** nemofac@naver.com / **www.**nemofactory.net

Collaboration / **SIMJI** bookstore
(135-010) 서울특별시 강남구 논현동 34-24 1층
T_ 02-543-3476 / 02-512-1788 / **F_** 02-543-3477 / **www.**simjibook.com _ simjibook1@naver.com

ⓒ NemoFactory / **ISBN** 979-11-88529-10-0(13540) / **Designed by** Joon-hee, Lee

SMALL CITY

Steel / Concrete / Glass / Wood / Soil / Container

HOUSE

HAZUKASHI-HOUSE ALTS DESIG
/ HOUSE IN MIYAMOTO TATO A
AKASAKA SHINICHIRO ATELIER /
ARCHITECTS / ARBOR HOUSE IN T
+ ASSOCIATES / EN-HOUSE MEC
/ HOUSE I YOSHICHIKA TAKAG
ARCHITECTS & ASSOCIATES / C
ARCHITECTS / OTSU HOUSE A
HOUSE FORM/KOUICHI KIMURA A
FUJIWARAMURO ARCHITECTS /
KIMURA ARCHITECTS / SLENDE
ARCHITECTS / TRANE APOLLO A
HOUSE HENNING STUMMEL ARC
AKI HAMADA ARCHITECTS / H
ARCHITECTS / ISHIBEHIGASHI H
APOLLO ARCHITECTS & ASSOCIA
ASSOCIATES / GARAGE HALL HOU
KEITARCHI / JUBAKO HOUSE MEC
PEAK APOLLO ARCHITECTS & ASS
GABLE ROOF HOUSE ALPHAVILLE
/ HOUSE FOR A PHOTOGRAPHER I
SCOPE APOLLO ARC

FFICE / **DODGED HOUSE** BUREAU
HITECTS / **STREAMLET HOUSE**
IKEDA YUKIE ONO TOSHIHARU
RUGASHIMA YOSHICHIKA TAKAGI
RO ARCHITECTURE LABORATORY
ASSOCIATES / **COVER** APOLLO
LTON HOUSE TOM ROBERTSON
S DESIGN OFFICE / **ADORABLE**
CHITECTS / **HOUSE IN TOYONAKA**
USE OF SCENES FORM /KOUICHI
HOUSE FORM/KOUICHI KIMURA
CHITECTS & ASSOCIATES / **GATE**
TECTS / **HOUSE IN SHIMOMURAKI**
SE IN MINOH FUJIWARAMURO
SE ALTS DESIGN OFFICE / **ACRO**
S / **GRID** APOLLO ARCHITECTS &
TSUKAGOSHI MIYASHITA SEKKEI +
RO ARCHITECTURE LABORATORY /
IATES / **HOUSE IN NIKKO** TAKIBI /
OLDED HOUSE PROCTOR & SHAW
RM/KOUICHI KIMURA ARCHITECTS /
TECTS & ASSOCIATES

Image _ Slender House _ FORM/Kouichi Kimura Architects

CONTENTS

012 **93.45** sq.m. **Hazukashi-House** / ALTS Design Office
020 **94** sq.m. **Dodged House** / BUREAU
034 **94.49** sq.m. **House in Miyamoto** / Tato Architects
046 **95.03** sq.m. **Streamlet House** / Akasaka Shinichiro Atelier
056 **96** sq.m. **MU** / Ikeda Yukie Ono Toshiharu Architects
066 **97.24** sq.m. **Arbor house in Tsurugashima** / Yoshichika Takagi + associates
074 **98.54** sq.m. **EN-House** / Meguro Architecture Laboratory
086 **100.24** sq.m. **House I** / Yoshichika Takagi + associates
096 **101.56** sq.m. **COVER** / APOLLO Architects & Associates
106 **104** sq.m. **Carlton House** / Tom Robertson Architects
118 **105.57** sq.m. **Otsu House** / ALTS Design Office
126 **106,4** sq.m. **Adorable House** / FORM/Kouichi Kimura Architects
134 **107.62** sq.m. **House in Toyonaka** / Fujiwaramuro Architects
142 **112.27** sq.m. **House of Scenes** / FORM/Kouichi Kimura Architects
154 **112.28** sq.m. **Slender House** / FORM/Kouichi Kimura Architects
164 **116.57** sq.m. **TRANE** / APOLLO Architects & Associates
176 **117** sq.m. **Gate House** / Henning Stummel Architects
186 **118.46** sq.m. **House in Shimomuraki** / Aki Hamada Architects
196 **119.46** sq.m. **House in Minoh** / Fujiwaramuro Architects
204 **122.90** sq.m. **Ishibehigashi House** / ALTS Design Office
214 **126.44** sq.m. **ACRO** / APOLLO Architects & Associates
222 **126.44** sq.m. **GRID** / APOLLO Architects & Associates
232 **131.04** sq.m. **Garage Hall House** / Tsukagoshi Miyashita Sekkei + Keitarchi
244 **138.99** sq.m. **Jubako House** / Meguro Architecture Laboratory
256 **143.95** sq.m. **PEAK** / APOLLO Architects & Associates
266 **152.22** sq.m. **House in Nikko** / TAKiBI
278 **157.77** sq.m. **Gable Roof House** / Alphaville
292 **162** sq.m. **Folded House** / Proctor & Shaw
302 **169,41** sq.m. **House for a Photographer** / FORM/Kouichi Kimura Architects
314 **180.41** sq.m. **SCOPE** / APOLLO Architects & Associates
323 **Participation** Designers

| **93.45**sq.m. | **Wood** |

Hazukashi-House

by ALTS Design Office

Location _ Fushimi-ku, Kyoto, Japan
Site Area _ 107.99m²
Building Area _ 53.46m²
Total Floor Area _ 93.45m²
Structure _ Wooden
Photography _ Yuta Yamada

13

This plan is the project that thought about how you take in rich light in a site condition of small space only in Kyoto. Is provided with a dining space gathering of family at the center of the house, that So we thought, was a blow-out there. This good space is also space which connects a family's bonds and also achieves dynamic functions, such as display and room of stairs... Also captures the light and wind diverse that while taking advantage of the material antique client demand, opening the rich opening, we have created a space with the charm which we want to put a face from anywhere in the house were.

Elevation

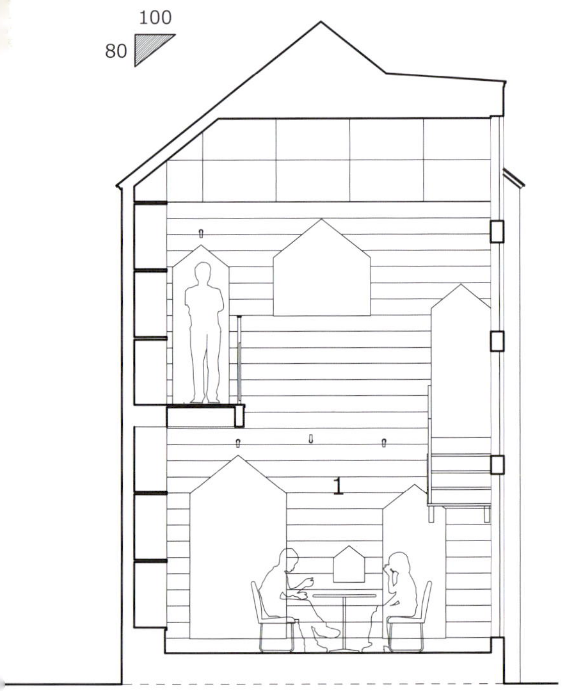

Section

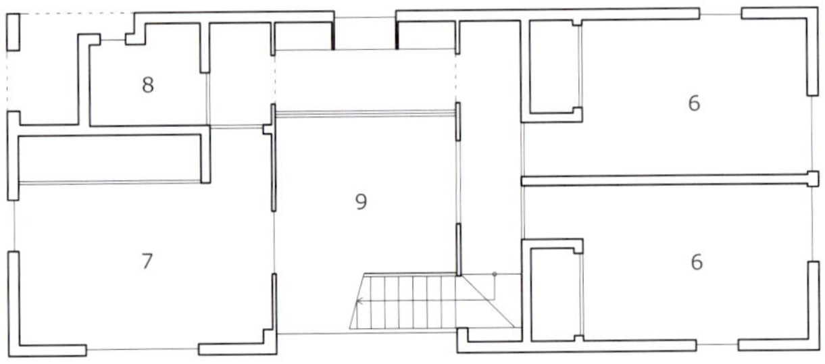

First Floor

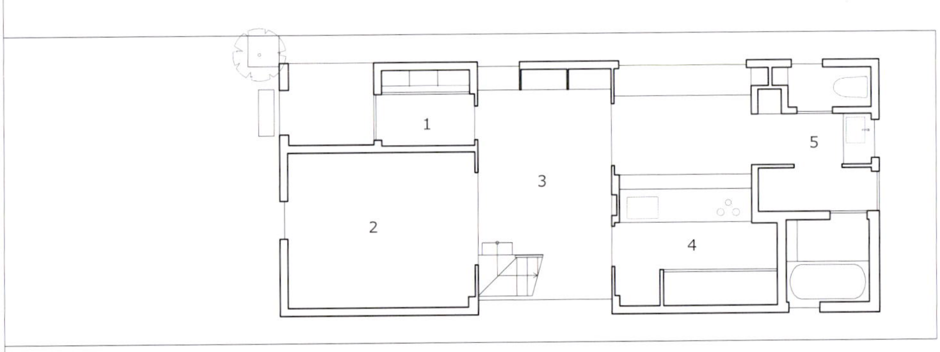

Ground Floor

1 : entrance 2:living 3:dining 4:kitchen
5:washroom 6:children's room 7:bedroom 8 : study 9 : void

| **94**sq.m. | **Concrete** |

20

Dodged House
by BUREAU

Location _ Lisbon, Portugal
Project _ Daniel Zamarbide (BUREAU) with Leopold Banchini Architects
Area_ $94m^2$
Photography _ Dylan Perrenoud

Image _ Irving Gill, Dodge House, California 1916, Picture by Marvin Rand

The crisis that hit Portugal ten years ago has produced an incredible density of abandoned spaces. The two main cities, Porto and Lisbon, offered a landscape of ruins and closed buildings that charmed an international community looking for a southern romanticism. Since then the two cities have acted and reacted to renew their historical centers and a good quantity of these abandoned houses have been renovated with general undeniable quality, probably due to the sensitive and cultivated approach of Portuguese architects in general. This landscape of closed buildings was featured by numerous opaque facades, hiding the interiors as if the life of those buildings had disappeared or was in a frozen state, waiting for better times to open the windows again and let the sunshine in. Streets with no windows face without eyes. This particular situation has evolved in Lisbon at a faster pace. The city center's reconstruction has opened the eyes to welcome an aggressive airbnb economy re-creating a well-known phenomenon that other cities in Europe have already gone through. The Air Bed & Breakfast also produces an Air Economy with implications and consequences that will fatally pop up with time. The impact of this market on the architecture realm is interesting. Renovations bear all a similar signature, creating qualitative and sensitive interventions, yet very marketable and instagramable, which seems to be the most important indicator of a successful architecture in this context.

South Elevation

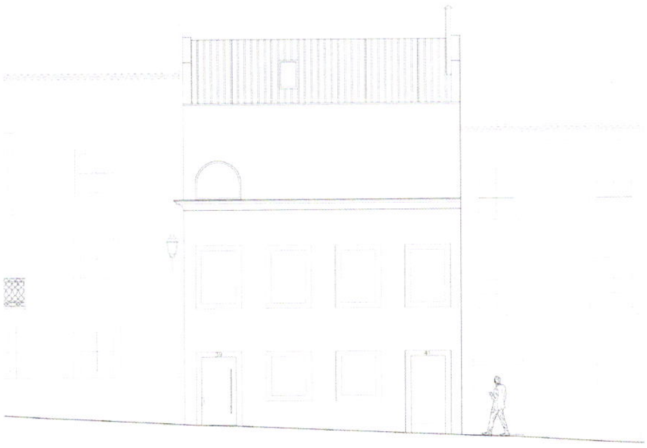

North Elevation

Two hundred years ago, at the turn of the 19th century, an architect looking for a healthier climate drove down from cold Chicago to California to start his practice in San Diego. He had been a draughtsman in the most influential office of the time, Sullivan's, and had actually worked under the direction of Frank Lloyd Wright's team for the now historical iconic Transport building. Irving Gill commenced then an astonishing creative career as an inventor and an understated avant-gardist. The story is interesting when comparing to the reach of media between these two moments of history. Today's immediacy of actions-images featuring Portuguese new renovated interiors is hardly comparable to the time of Gill; his architecture had to wait until Esther McCoy wrote the now well-known publication Five California Architects in 1960 to be acknowledged and celebrated. It was actually Reyner Banham some years later who showed this work to the international public in his book, Los Angeles: The Architecture of Four Ecologies in 1971. Despite this late success the destruction of one of his main works, The Dodge House, could not be avoided, even though a strong campaign that Esther McCoy carried out to save this historical icon of Californian architecture.

The Dodged House in Lisbon pays a double tribute. On one hand to Irving Gill's architecture. The very particular modernity that he established as the basis of his practice seems to perfectly echo the Portuguese context (the same way as Gill's architecture was understood to develop from the Missions in California). On the other hand, as a trace of the time in which the Dodged house was designed and built, it has preferred to keep her eyes closed and opaque façade and has bet on a less marketable feature, space, void, interior volume that refuses efficiency of land use. Within a rather small plot (around 40 m2 ground floor, 94 m2 in total) the Dodged house has privileged a string section and a contemplative void, proposing a diversity of interior-exterior spaces that extend into a courtyard.

Evidently, the project responds as well to complexity of functional requirements that has turned the house into a "machine à habiter", playing again, quite deliberately and strongly with the history of modernism and its inhabitable typologies. In the end, the Dodged house is quite a simple and readable project. Although it might be complex in its inscription into the urban fabric and historical context, it is nevertheless quite straightforward in its way to occupy space and distribute the program in a small plot. As its names indicate, the Dodged house makes an attempt to elude, to trick, an actual state of a certain architecture in Lisbon.

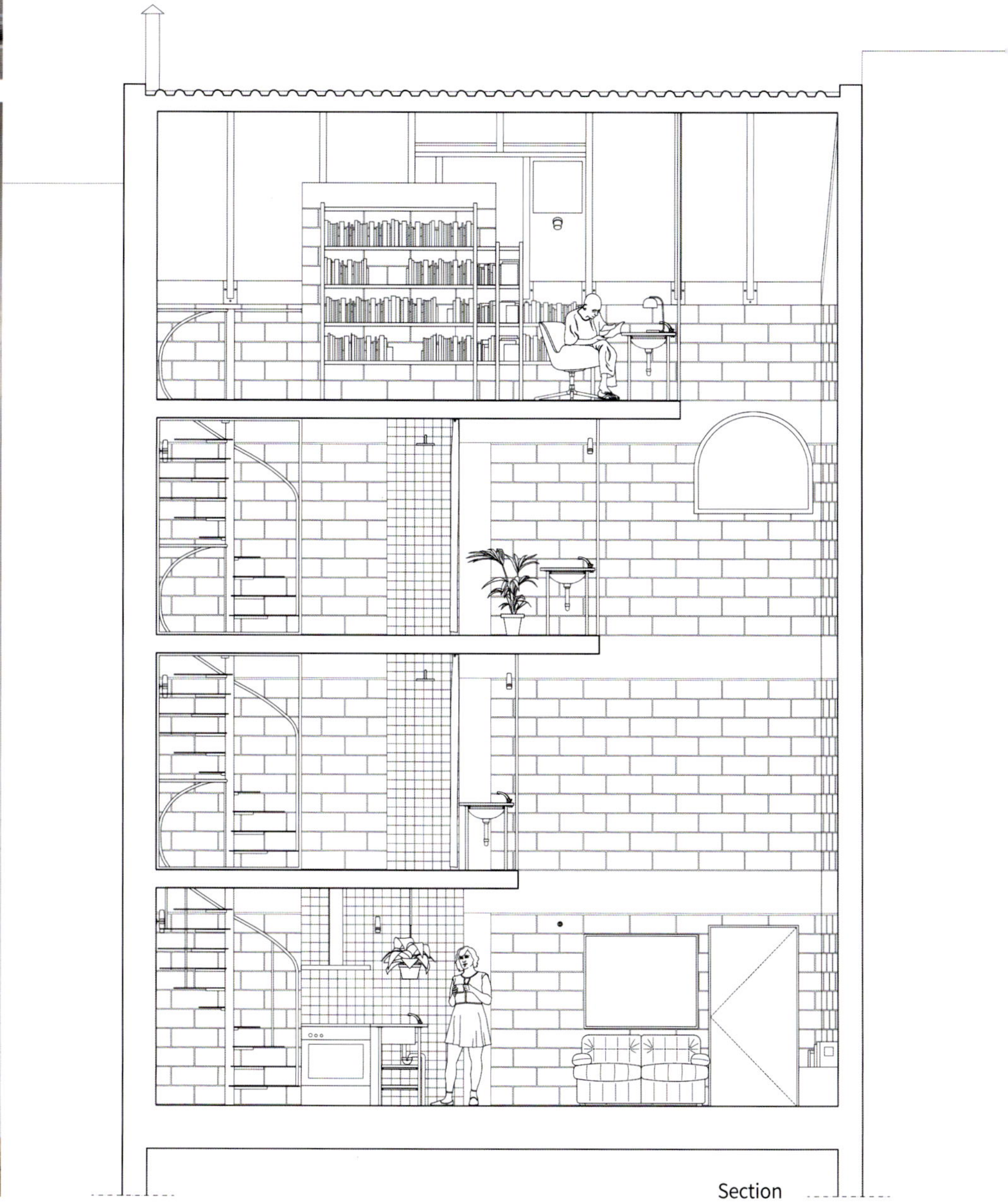

Section

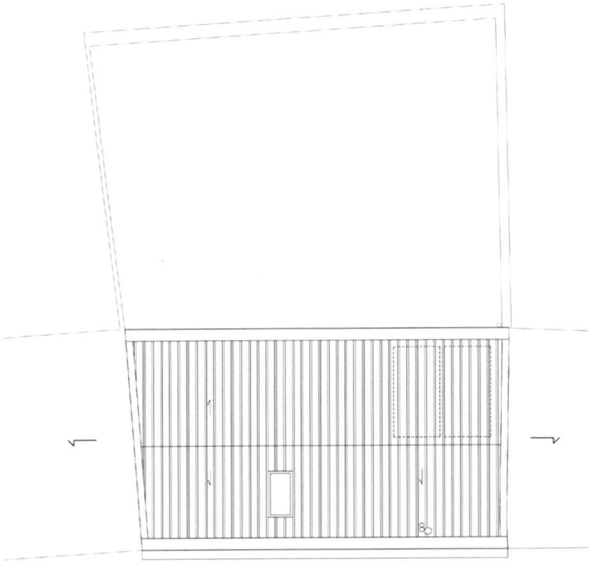

Roof Floor

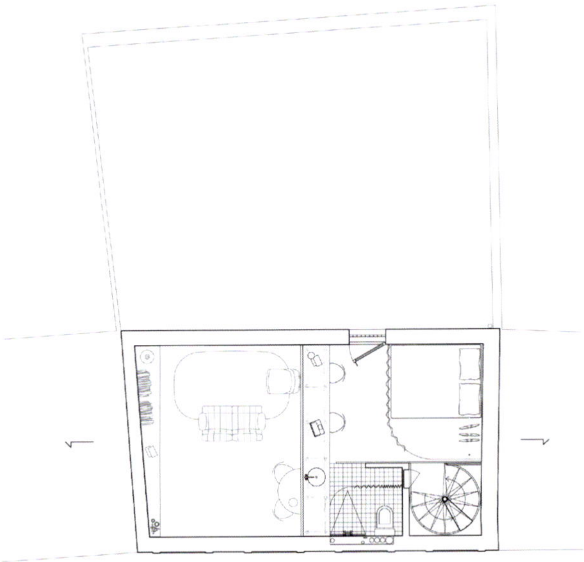

First Floor

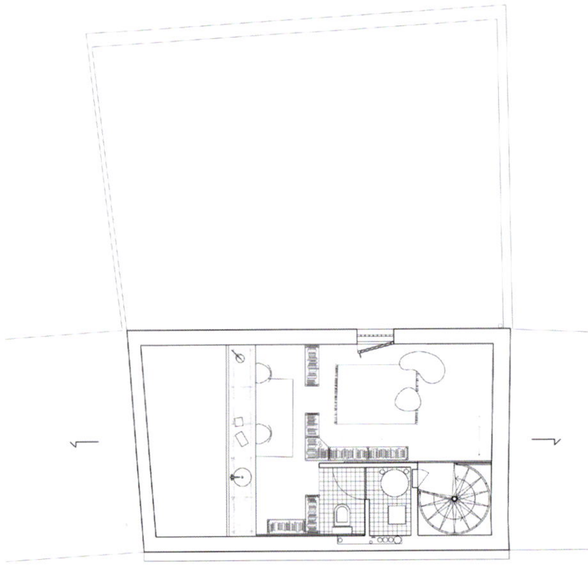

Third Floor

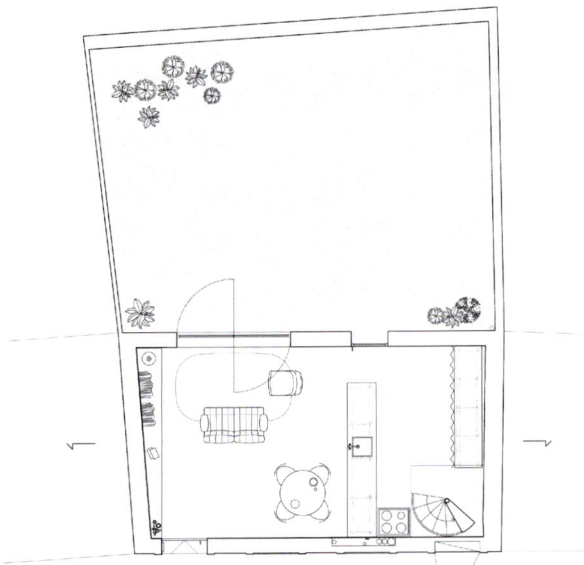

Ground Floor

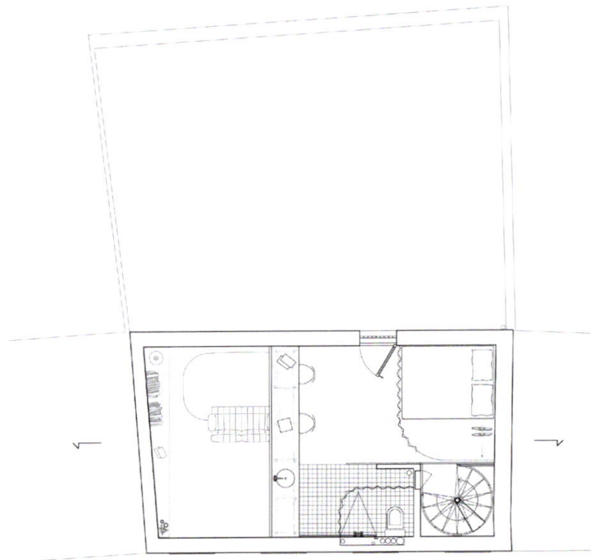

Second Floor

| **94.49**sq.m. | **Steel** |

House in Miyamoto

by Tato Architects

Location _ Osaka, Japan
Design Team _ Yo Shimada, Nobuhiko Sato, Koji Hoshiyasu
Site Area _ 128.19m^2
Building Area _ 49.73m^2
Total Floor Area _ 94.49m^2
Building Scale _ Two-storey House
Structure _ Steel Flame
Construction _ Construction
Structure Design _ Takashi Manda Structural Design
Photography _ shinkenchiku_sha

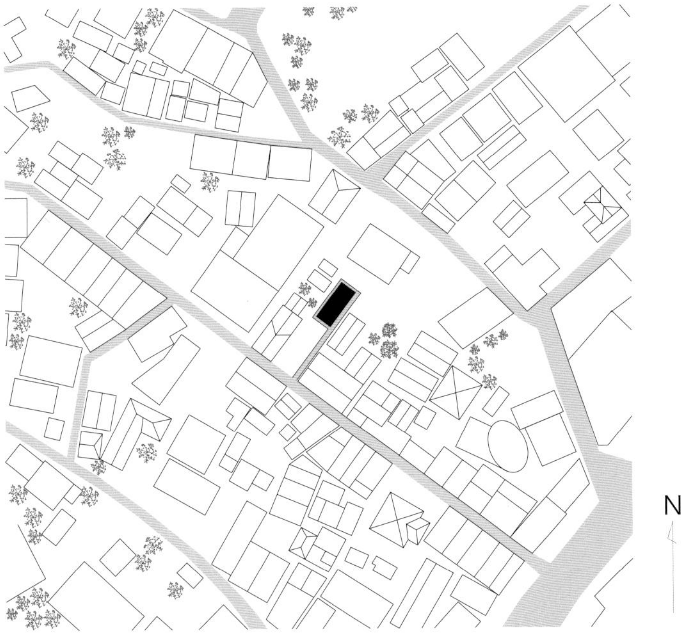

At this house - a residence designed for a family of three and their many belongings - the client requested that the members of the family can feel close to each other regardless of where they are in the house. Moreover, private rooms were not needed because they feel that it is lonely to withdraw into one's space, and storage space was also unnecessary because they did not want to tuck things away. As a result of trying to find a form that allows the whole house to feel like one room while securing sufficient space for their belongings at the same time, we proposed a design to connect the flooring with a height difference of 700 mm, where the different levels can be used as tables and shelves. The floors build up as two spiral shapes, joins at the living room, and then separate into two again before arriving at the rooftop deck. By using this combination of two spirals, we were able to create multiple paths inside the house that allows different room compartments and changes in the circulation, equipping the house to be able to accommodate changes in the lifestyle of the client.

The surrounding of the flag-shaped site as a result of the selling of the old wooden house near the station, are parking lots and apartments, and it is expected that there will be more tall buildings built in the near future. Assuming that the only outdoor area that receives good sunlight is the roof, a rectangular box-shaped volume with the flat roof and triangular terraces are inserted, the windows are placed uniformly with the least interference with the structure to allow the house to be able to accommodate changes in the environment. Due to the fact that the house is located in a fire prevention district, it is designed with a steel framework with a ceiling height of 6,900 mm, where 13 floors surface float, with the top seven different levels suspended with 20 mm steel rods from the roof beam, with the bottom six floors supported by 75 mm square steel pipes. The spatial structure that is constructed by repeating a simple autonomous system is similar to an "echo chamber", which amplifies the innermost lifestyle of the client and conveys a sensitivity that expands without limit.

The client has lived in the area nearby and slowly moved into the house. Since they started officially living here, the things, the architecture, and people has become one like an ensemble, and the rich scenery of their lifestyle is expanding as if they are moving inside a forest. From now on, we look forward to seeing the house the things, the residents and architecture settle down in relation to each other through time.

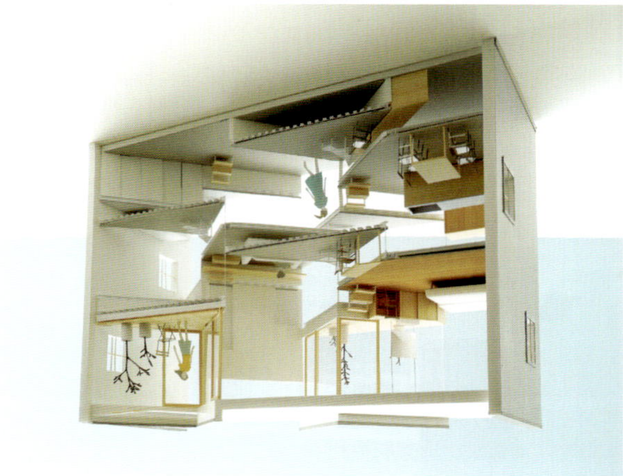

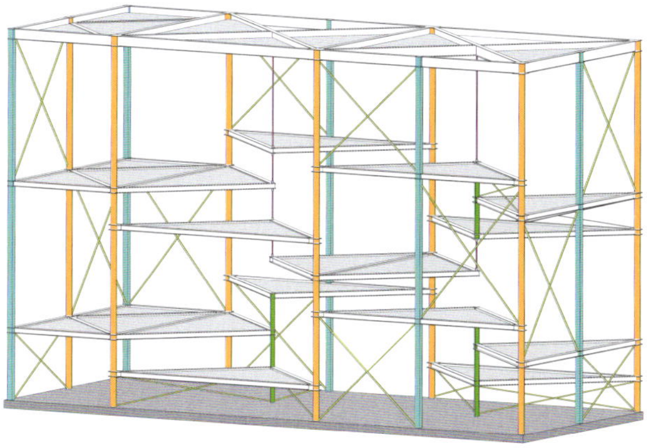

凡例
- 柱 □ 100×100×9.0mm STKR400
- 柱 ○ 11.43×6.0mm STK400
- 柱 □ 75×75×4.5mm STKR400
- ブレース M20(TB) SNR400B
- 吊り材 M20(TB) SNR400B

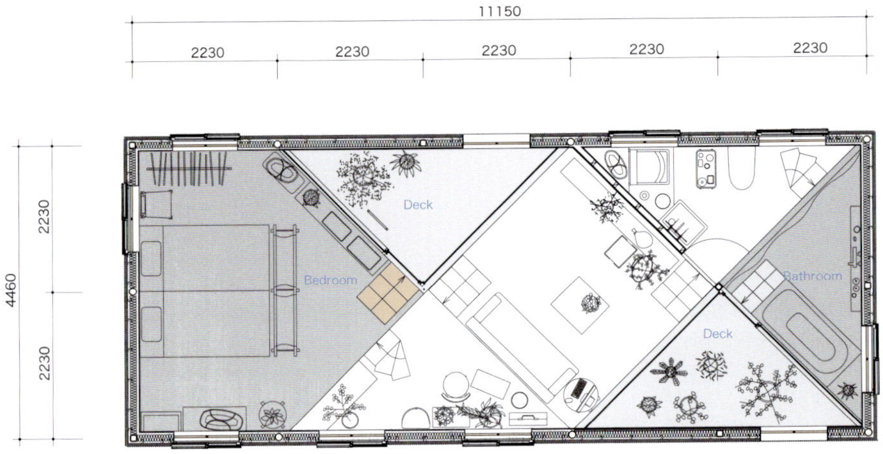

GL+5220

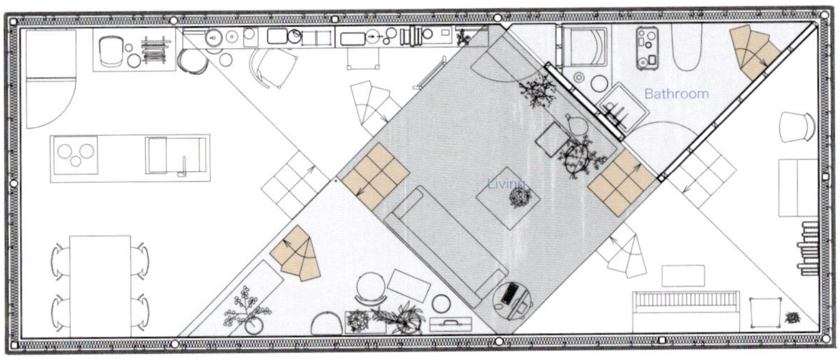

GL+3650

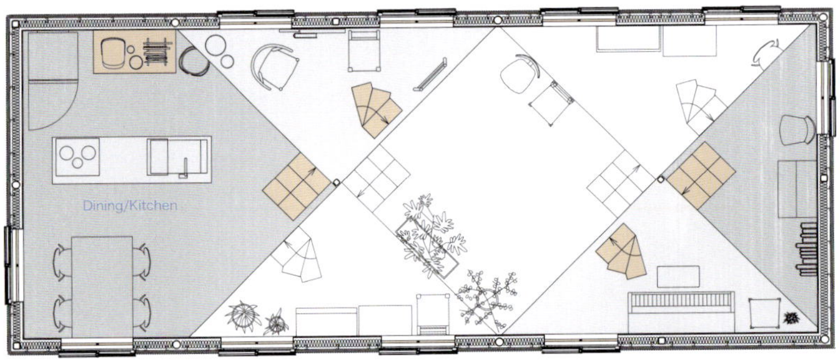

GL+2250

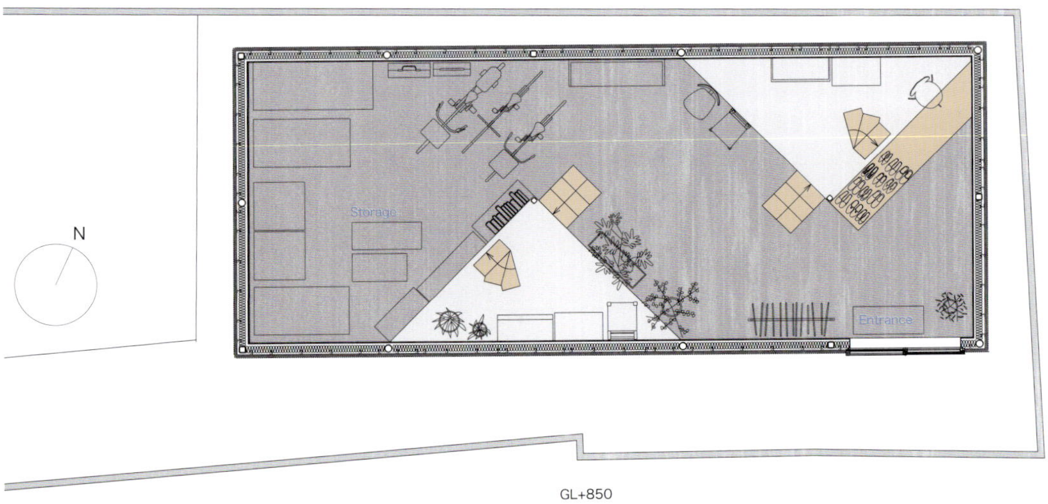

GL+850

| **95.03**sq.m. | **Wood** |

Streamlet House

by Akasaka Shinichiro Atelier

Location _ Sapporo, Hokkaido, Japan
Architect _ Shinichiro Akasaka / Hikari Sano / Yumiko Imasaka
Site Area _ 111.52m²
Building Area _ 46.23m²
Total Floor Area _ 95.03m²
Photography _ Koji Sakai

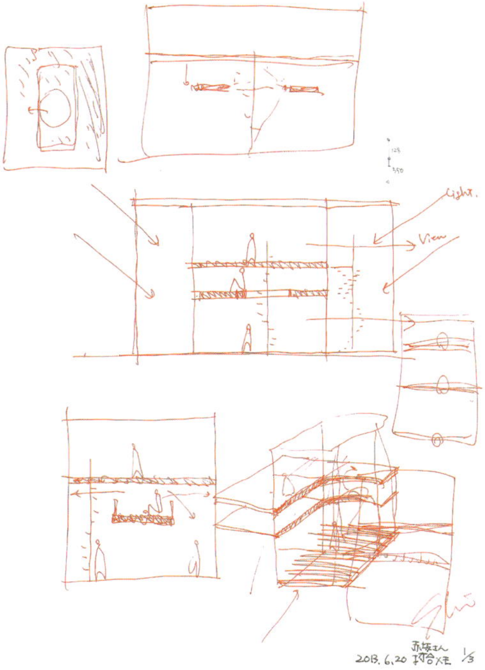

A long time ago in the city center of Sapporo, a spring pond called "Mem" and a clear stream meandering across through the city. However, as the city grows, streams were buried underground, and it is now difficult to track its traces. "Streamlet House" is built along the trace of the erased stream. While being covered with a concrete lid, this slightly arced "River land" crossing the grid-like residential dense area, is the approach of this flagpole site.

Along with the current dried river land, considering the former river edge, the entrance hall is placed with a potted tree planted, to add some moisture and also the green to continue to the interior. This hall vertically penetrates the entire house with five different floor levels. The scent of trees and the presence of family fill its high ceiling, to unite this dwelling into one. On the other hand, the roof terrace which can be easily approached from the interior via catwalk is designed to be filled with green using light-soils in the future. The high-side lights are installed to look at flowers on rooftop swaying in the wind, which is also used as wind-vent and lighting window. Although the number of openings facing the outside is limited due to the location surrounded by apartments and houses on all sides, this house is aimed to meet the nature hidden in clutter of city center, by using the motion and coloring of green arranged on the opposite side from the several openings which were positioned carefully.

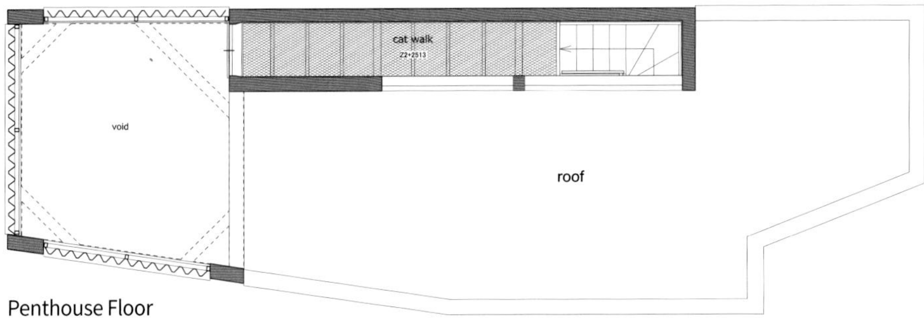

Penthouse Floor

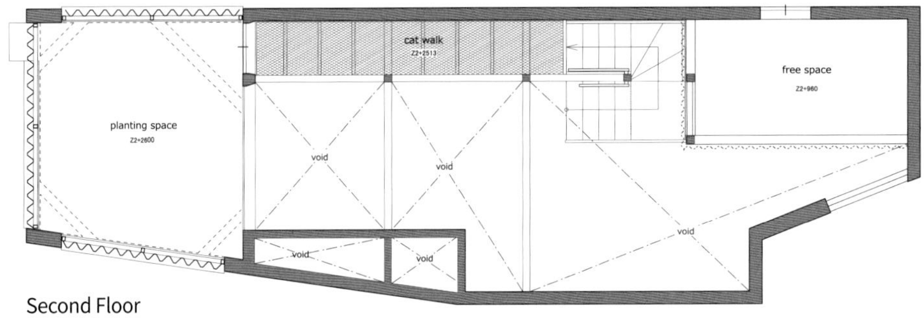

Second Floor

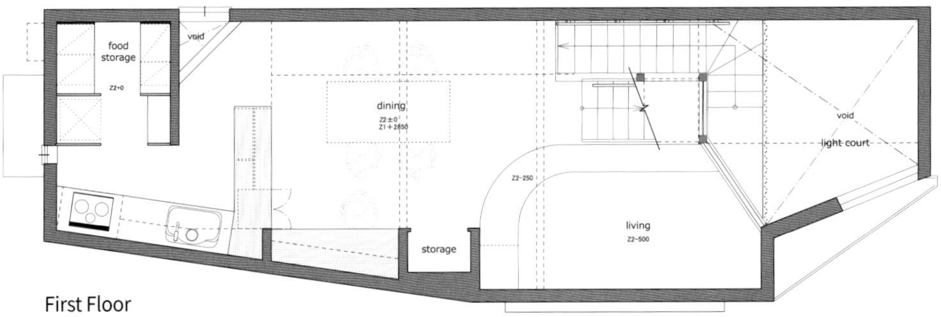

First Floor

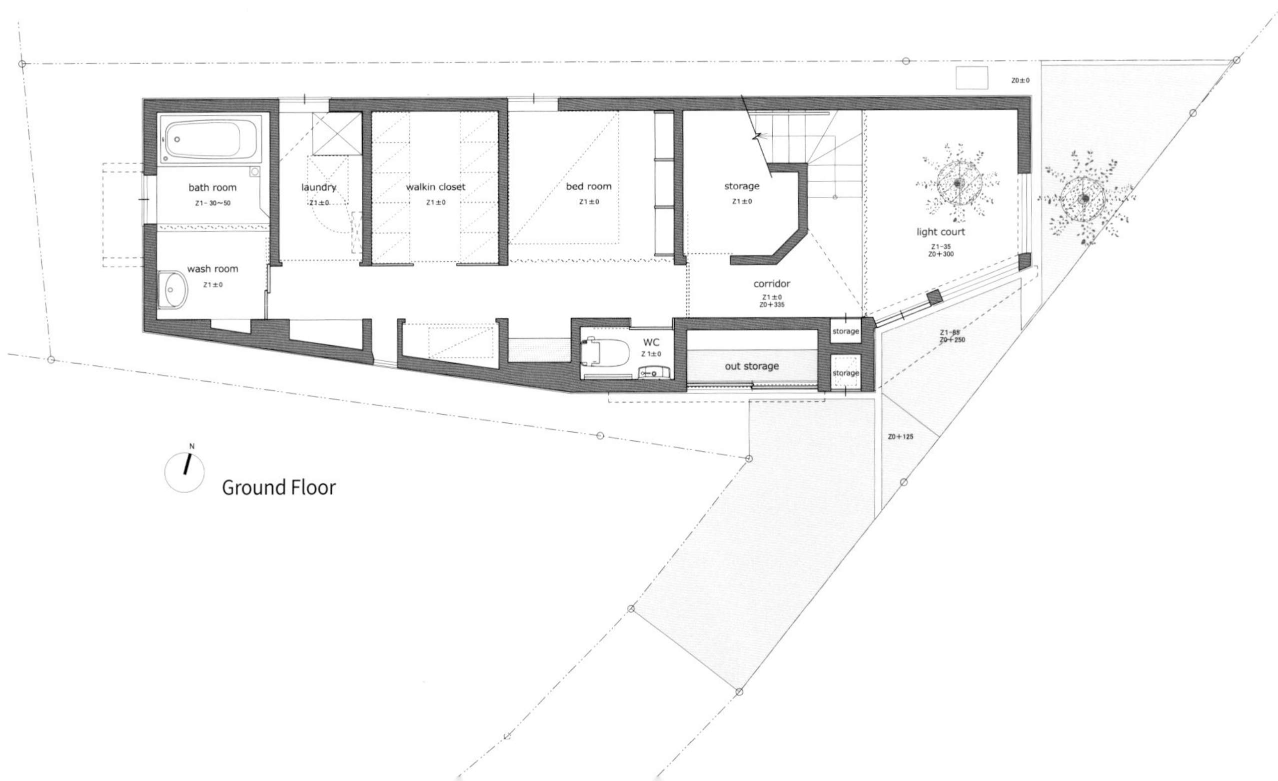

Ground Floor

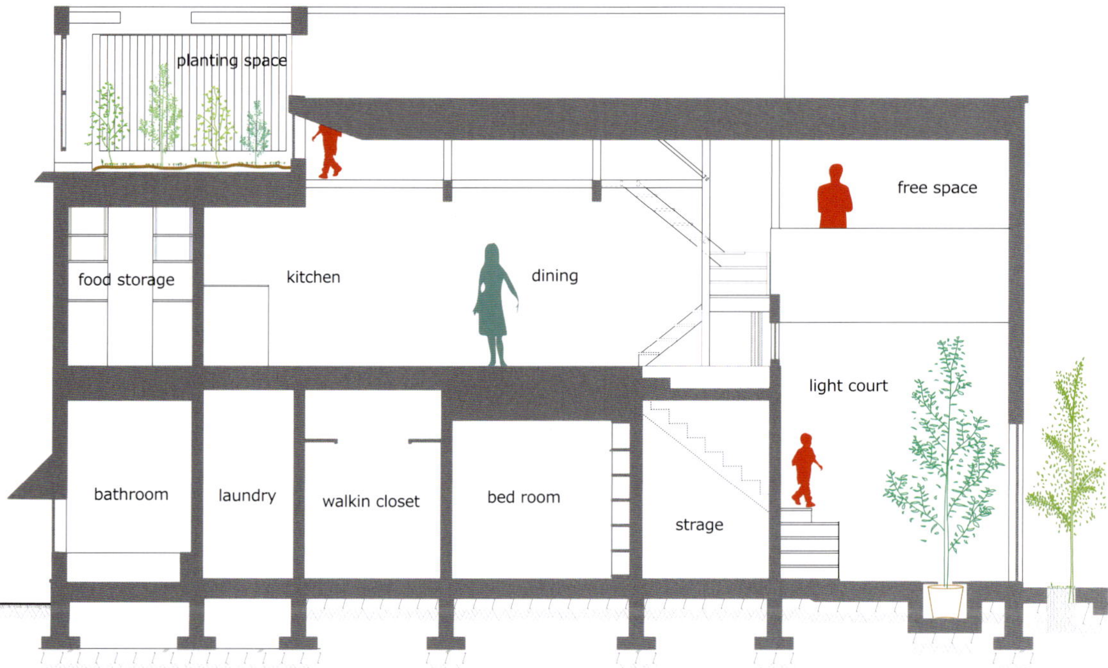
planting space
food storage
kitchen
dining
free space
light court
bathroom
laundry
walkin closet
bed room
strage

Section

| **96**sq.m. | **Wood** |

MU
by Ikeda Yukie Ono Toshiharu Architects

Location _ Hachinohe, Aomori, Japan
Floor Area _ 96m^2
Lead Architects _ Ikeda Yukie, Ono Toshiharu
Structural Engineer _ MID architectural structure laboratory
Constructor _ Tanabu Group
Photography _ koichi torimura

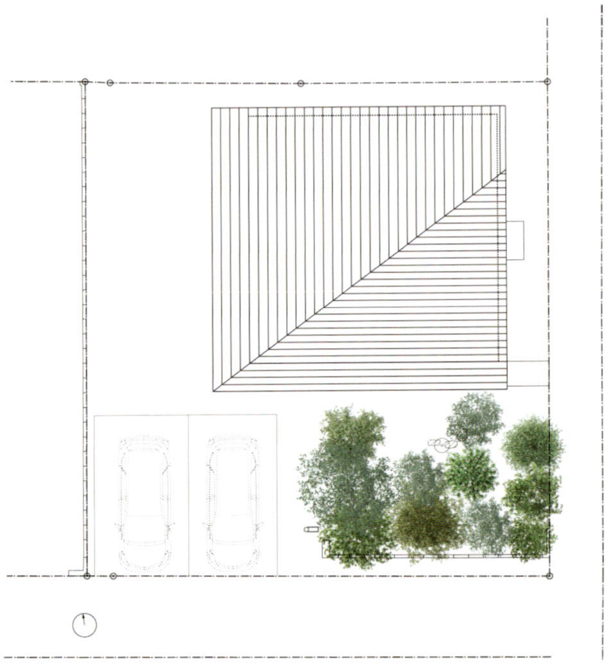

MU consists of a dwelling for a couple combined with a photo studio in the small northern city, Hachinohe, Aomori. People often gathered at the existing studio as if it were a meeting place of the town, and we aimed to maintain that atmosphere while considering the couple's privacy.

Taking the hard climate into account, we looked for openness, not only by physical transparency. In accordance with the low-rise residential neighborhood, the volume is formed into a deconstructed quadrangular pyramid with a low façade facing the street and rising back up towards the North. Inside the pyramid, the privacy-cores are piled in tiers, while the surrounding spaces are open. The mountain-like presence of the form invites people, while the large open space enclosed by powerful frame structures is receptive to both privacy and publicity. A privacy-core opens with sliding doors, which enables efficient use in the flexibility to adjust the area of private and public.

With this system, the dwelling can open a lot of the space to the public, and by that the dwellers in this shrinking city can gain social capital through rich relations.

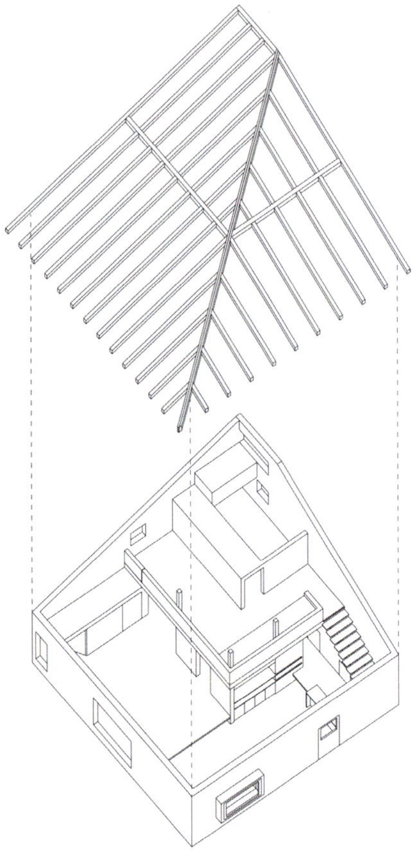

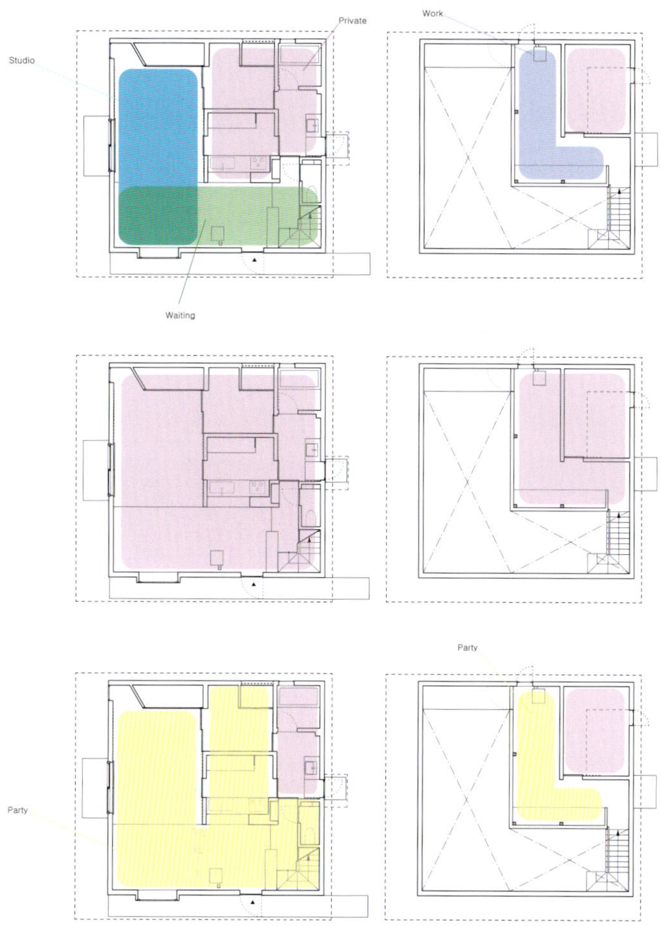

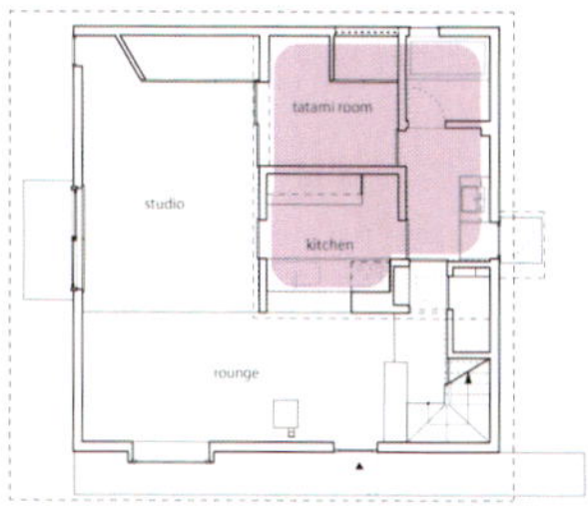

plan1F

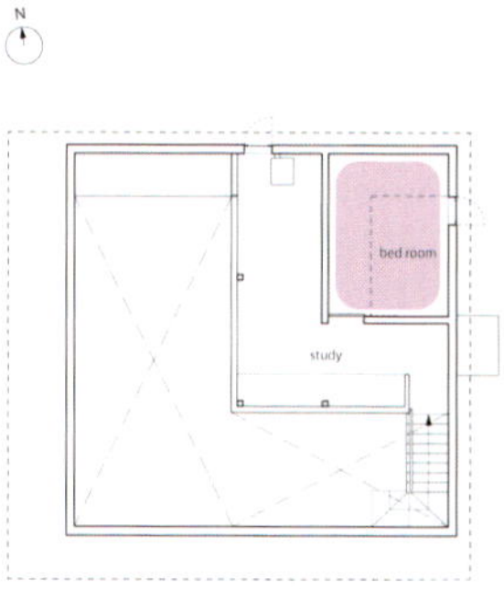

plan2F

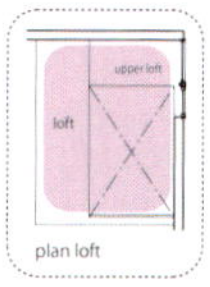

: private core

| **97.24**sq.m. | **Wood** |

Arbor house in Tsurugashima

by Yoshichika Takagi + associates

Location _ Tsurugashima, Japan
Structural Engineer _ Daisuke Hasegawa
Building Area _ 57.50m^2
Total Floor Area _ 97.24m^2
Manufacturers _ Lixil Corporation, Rinnai, YOSHINO, Sanwa company
Photography _ Yuta Oseto

This is a remodeling of a wooden house in an ordinary dense residential area in Tsurugashima, Kanto region. It was built 33 years ago on a backland site that connects to the road via access of only 2 meters wide. Before remodeling, it had a small bright garden on the south side. There was a field in the neighboring land that faces the garden, where the open sky could be seen. However, it was planned to be remodeled for residential development. Therefore it could be imagined that it would turn to a garden surrounded by houses, which is never a good environment.

The plan of this house consists of a combination of remodeling and extension. In the original plan, the house was modeled with separate living, kitchen and dining rooms. Since this had a closed atmosphere, we changed the plan and remodel the house with an open atmosphere. Shareable rooms such as a living room and a dining room now have open ceilings with vertical expanses of space, where working space and a loft were made. In addition, we decided to build a little arbor in the garden. A simple arbor has been built next to the living room in the remodeled house so that it would be consecutive with the living room spatially and visually.

The roof of the arbor is made of permeable material so that it can catch small light and then spread it. In the middle of summer, the arbor softens the intense sunlight like a lighting cover, creating proper brightness. It also plays the role of a window treatment so that the inside of the house cannot be seen by neighbors. If you open the windows of the living room, making the arbor consecutive with space, it will expand its functionality as a living room. The arbor will become a part of the living room, where you can enjoy reading books, eating meals and playing with your dog. It can also be used as a smoking room, a relaxing space which you can use every day.

The small arbor has changed the garden, becoming an important existence that influences the whole house.

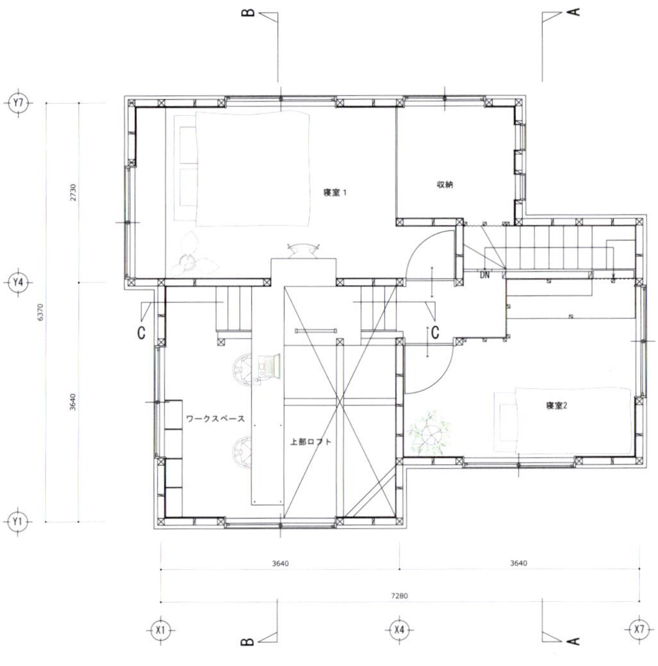

First Floor

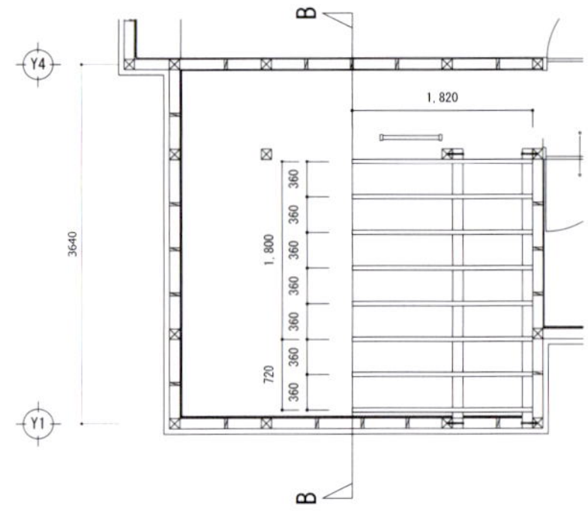

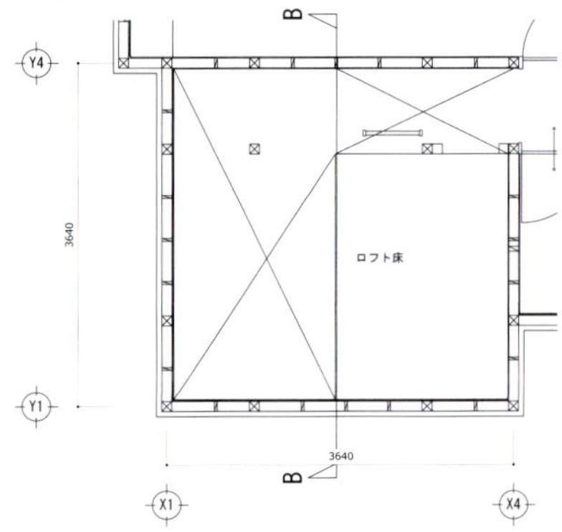

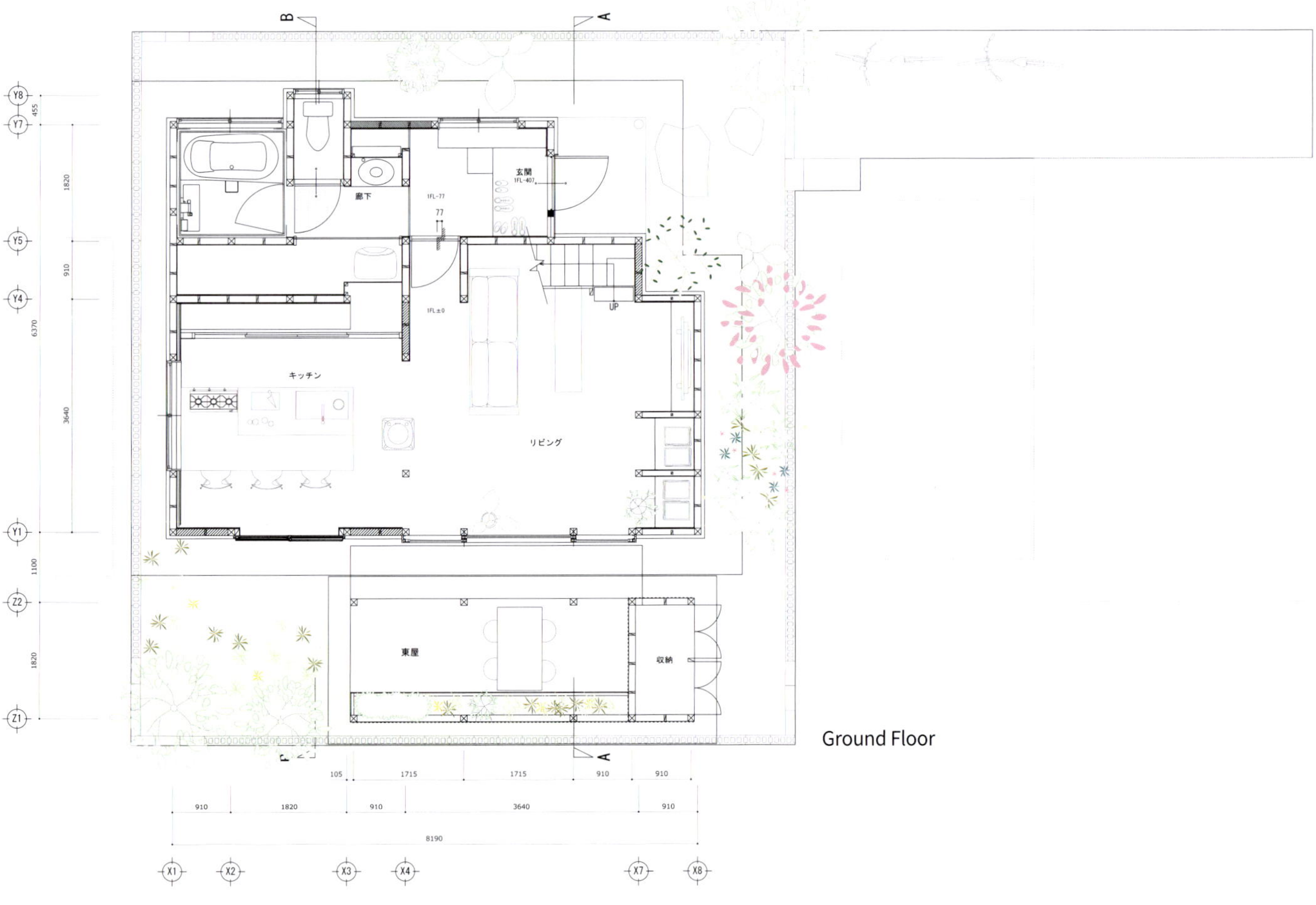

Ground Floor

| **98.54**sq.m. | **Timber** |

EN-House

by Meguro Architecture Laboratory

Location _ Tokyo, Japan
Construction nature _ Timber structures
Site Area_ 172.70m²
Building Area _ 57.13m²
Total Floor Area _ 98.54m²
Photography _ Koichi Torimura, Shinkenchiku-sha

Meguro Architecture Laboratory has completed an experimental residence that can be opened out liberally onto the town by reducing construction through renovation.

'En' in Japanese connects to the outside world from the occurrence of phenomena such as omens and fate and it is in this that there is significance that expresses the relationship in which change occurs. This expressed the structure and framework of the world with the word relationship. The 'en' in this residence is exactly this kind of place and is a platform for creating various lifestyle landscapes.

The grounds are in Teramachi and it is possible to feel the elegance of Edo from the townscape. However, in recent years many residences with high walls have been built facing on to the road and the residences and the town are detached from each other and have become closed off. The existing residence for which work was planned this time was surrounded by high walls in the same way and furthermore, it was necessary to backspace the south of the grounds by widening the front road as a legal requirement.

First of all, the structure was reduced to a skeleton then it was all updated to current earthquake-resistant standards by reinforcing the foundations and reinforcing weak earthquake resistance. While doing this, a sash was inserted after backspacing the existing exterior wall on the boundary facing the south road on the 1st floor by 910mm. The backspaced area was covered with wood decking and while this ensured an expansion of garden space, the interior and exterior boundaries were distanced from the road. The wood decking area is called ENGAWA and was determined as an interim area that connects the garden with the interior space. ENGAWA is an exterior location enveloped in calm due to the existing oblique exterior wall.

■PLAN DIAGRAM

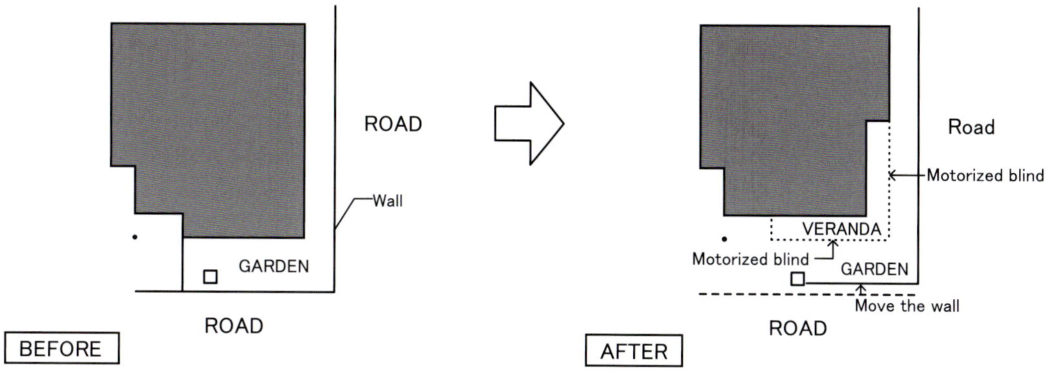

■SECTION DIAGRAM

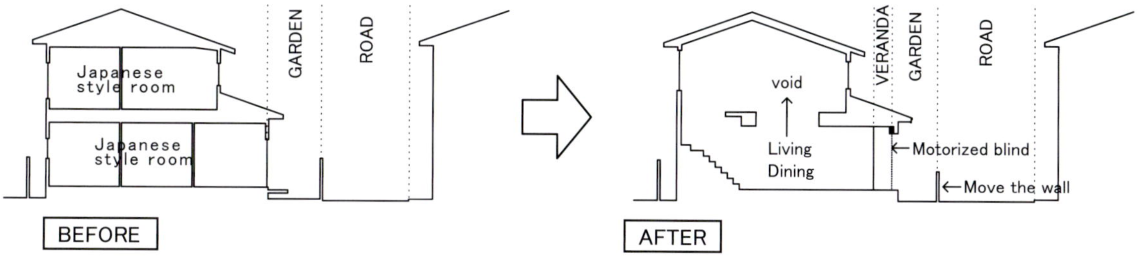

It has been made possible to convert the interior and exterior portions by installing an external blind in the original wall portion. Through these kinds of multi-layered relationships, even though the distance between the road and the building is now closer, the height of the line of sight can be controlled by the backspaced wall and the small garden has been transformed into a new space that connects the residence with the town.

The interior is a space in which you can experience the connection with the outside as is. A round atrium has been installed in the centre of the house in order to create a sense of unity between the upper and lower floors. (The pronunciation for 'round' in Japanese is also 'en'.) This gently cuts an arc for various lifestyle scenarios and creates new landscapes.

Beyond the small 2nd floor window that faces the atrium are the bedrooms and it is possible to make yourself heard in them from the 1st floor.

The large fixed fittings window in the gallery depicts the look of the sky on one side from the 1st floor kitchen past the atrium and when you ascend to the 2nd floor, the forest bursts in from beyond the town. In this way, people are connected with far away landscapes through buildings.

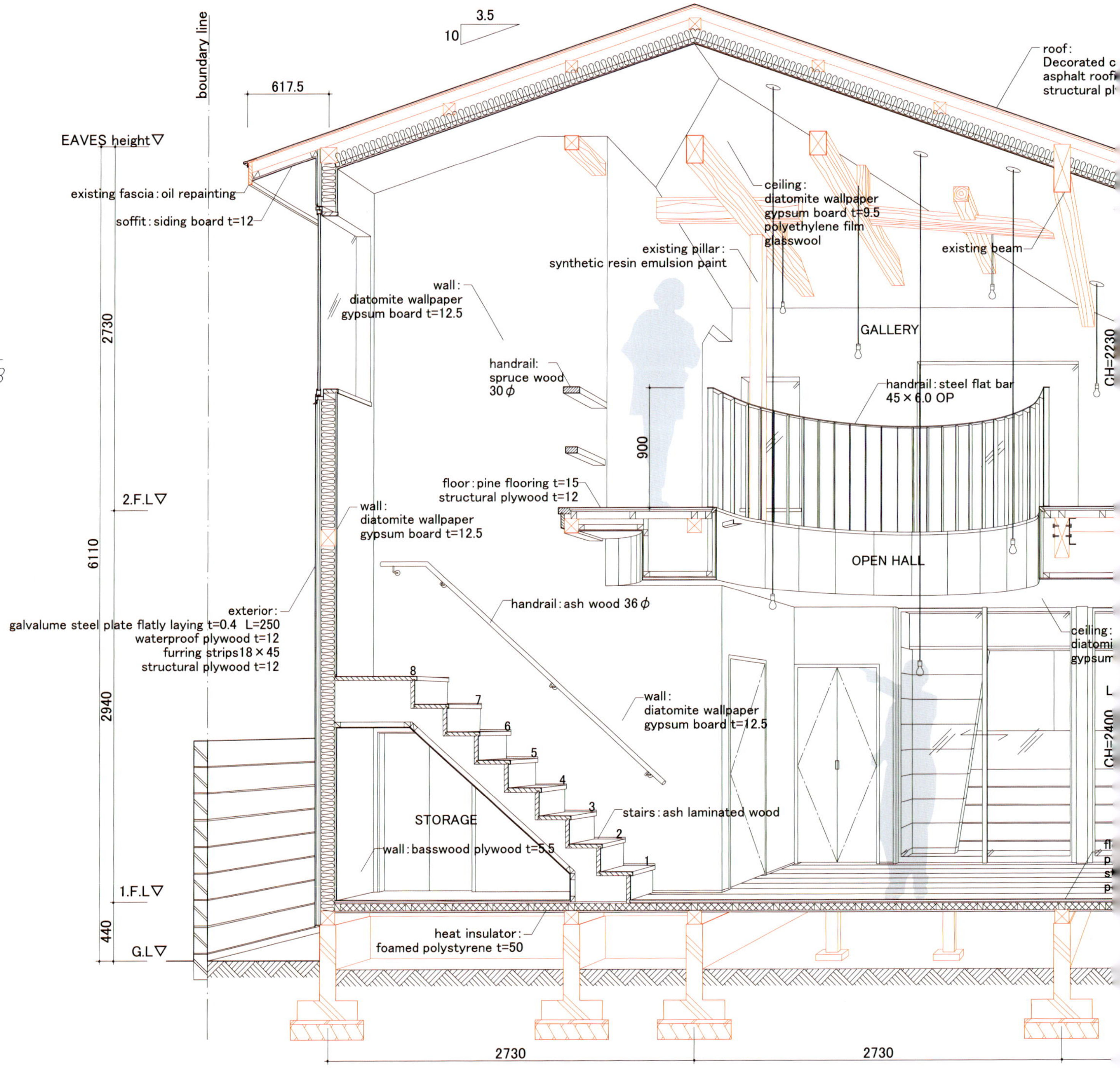

boundary line
3.5
10
617.5
roof:
Decorated c
asphalt roofi
structural pl
EAVES height▽
existing fascia: oil repainting
soffit: siding board t=12
ceiling:
diatomite wallpaper
gypsum board t=9.5
polyethylene film
glasswool
existing pillar:
synthetic resin emulsion paint
existing beam
wall:
diatomite wallpaper
gypsum board t=12.5
2730
GALLERY
CH=2230
handrail:
spruce wood
30 ϕ
handrail: steel flat bar
45 × 6.0 OP
900
floor: pine flooring t=15
structural plywood t=12
2.F.L▽
wall:
diatomite wallpaper
gypsum board t=12.5
6110
OPEN HALL
handrail: ash wood 36 ϕ
exterior:
galvalume steel plate flatly laying t=0.4 L=250
waterproof plywood t=12
furring strips18 × 45
structural plywood t=12
ceiling:
diatomi
gypsum
wall:
diatomite wallpaper
gypsum board t=12.5
2940
CH=2400
STORAGE
stairs: ash laminated wood
wall: basswood plywood t=5.5
1.F.L▽
heat insulator:
foamed polystyrene t=50
440
G.L▽
2730
2730

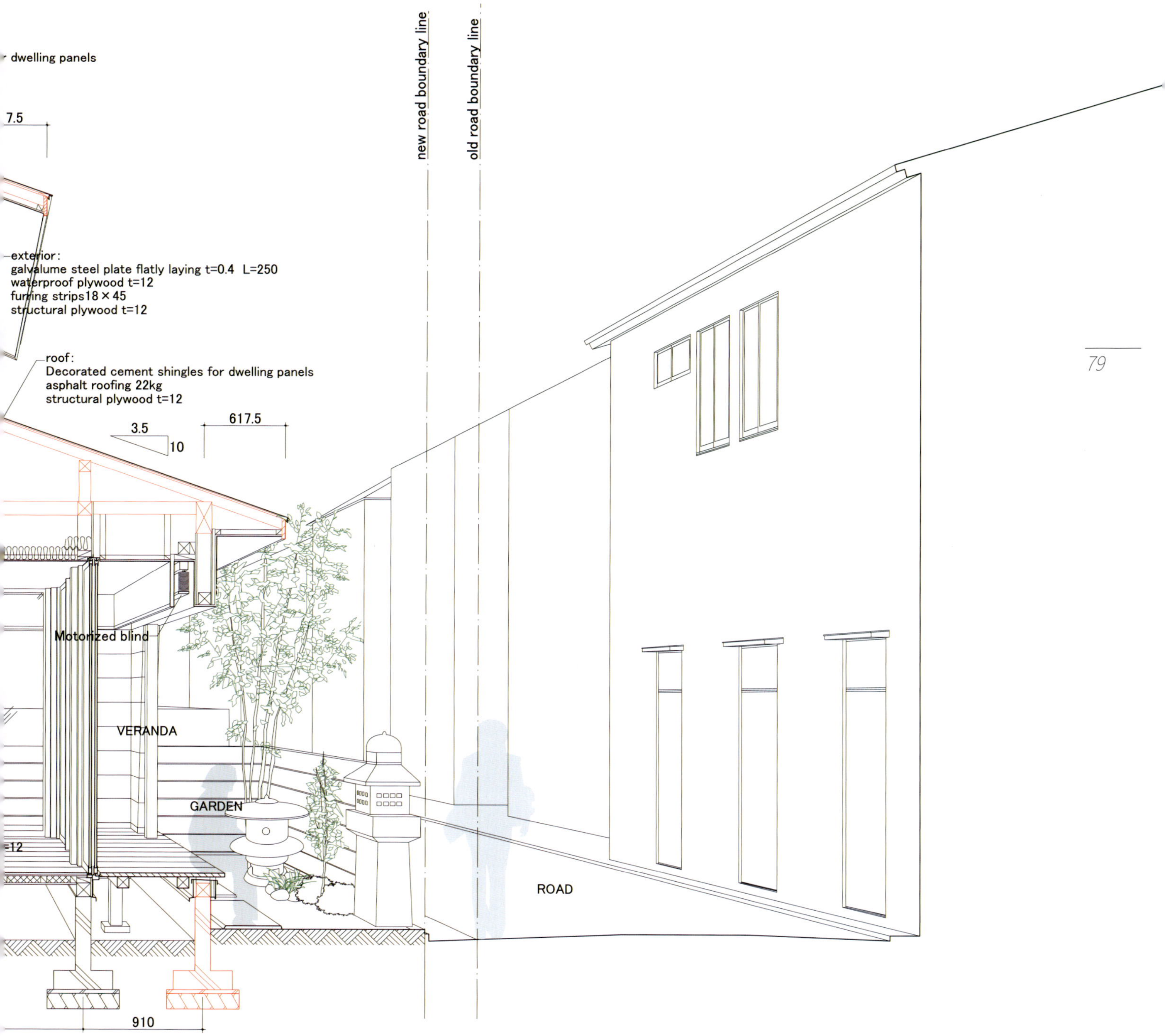
dwelling panels
7.5
new road boundary line
old road boundary line
exterior:
galvalume steel plate flatly laying t=0.4 L=250
waterproof plywood t=12
furring strips18 × 45
structural plywood t=12
roof:
Decorated cement shingles for dwelling panels
asphalt roofing 22kg
structural plywood t=12
3.5
10
617.5
Motorized blind
VERANDA
GARDEN
=12
ROAD
910

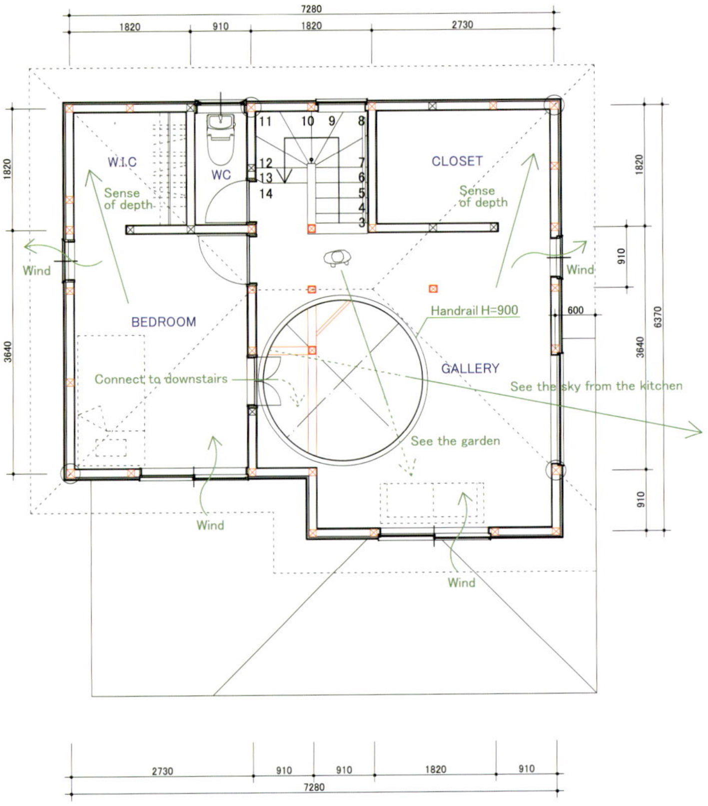
W.I.C
WC
CLOSET
Sense of depth
Sense of depth
Wind
Wind
BEDROOM
Handrail H=900
GALLERY
Connect to downstairs
See the sky from the kitchen
See the garden
Wind
Wind

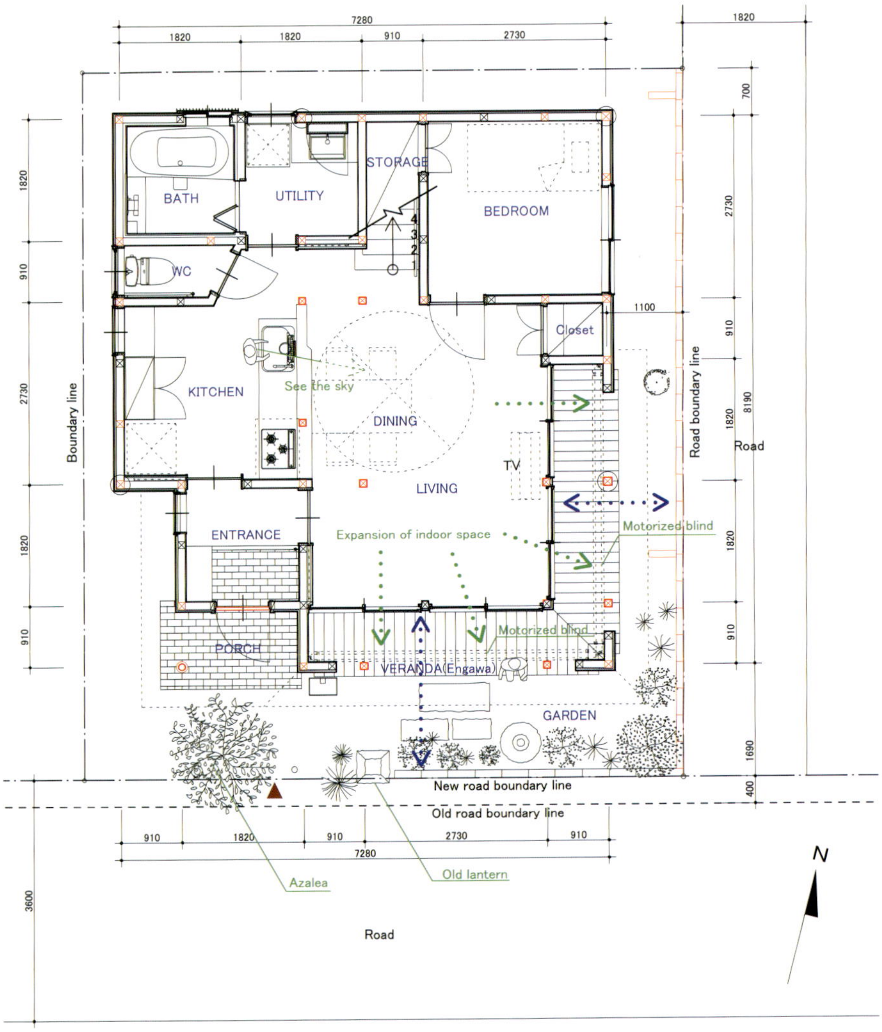
STORAGE
BATH
UTILITY
BEDROOM
WC
Closet
KITCHEN
See the sky
DINING
TV
LIVING
Boundary line
Road boundary line
Road
Motorized blind
ENTRANCE
Expansion of indoor space
PORCH
VERANDA(Engawa)
GARDEN
New road boundary line
Old road boundary line
Azalea
Old lantern
Road
N

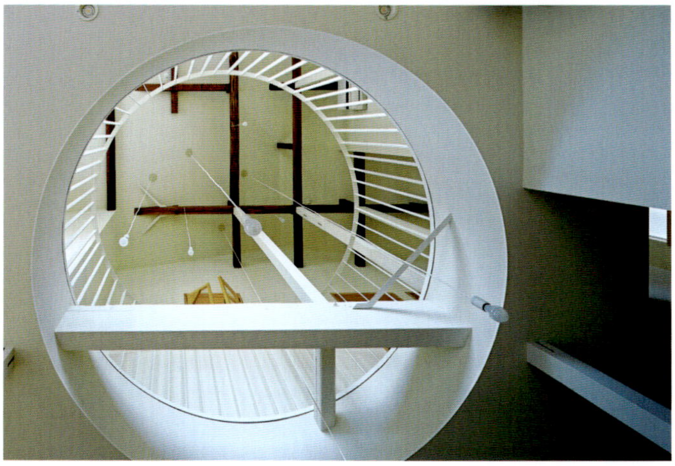

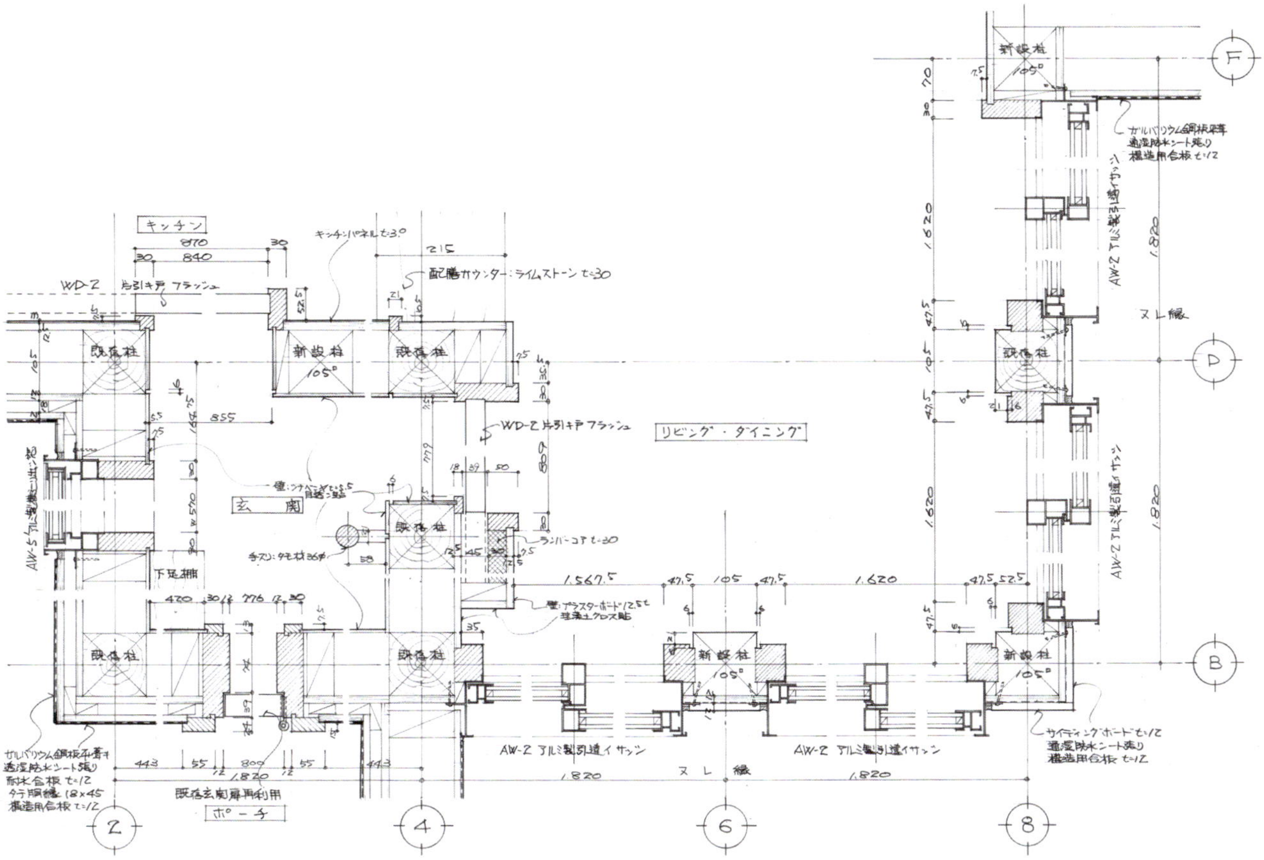
キッチン
玄関
リビング・ダイニング
ポーチ
既存玄関扉再利用
下足棚
新設柱
既存柱
WD-Z 片引キ戸 フラッシュ
AW-Z アルミ製引違イサッシ
AW-5 アルミ製縦すべり出し窓
ヌレ縁
配膳カウンター：ライムストーン t=30
キッチンパネル t=3.0
サイディングボード t=12
透湿防水シート張り
構造用合板 t=12
ガルバリウム鋼板平葺キ
1,820
1,620
2

| **100.24**sq.m. | **Wood** |

86

House I

by Yoshichika Takagi + associates

Location _ Akita, Japan
Structural Design _ Daisuke Hasegawa (Daisuke Hasegawa & Partners)
Gross Useable Floor Space _ 100.24m²
Lot Size _ 476.99m²
Structure _ Wooden Structure
Walls Surfaces _ Garvanised Stealseat Siding
Kitchen Companies _ Designed by Yoshichika Takagi
Photography _ Toshiyuki Yano

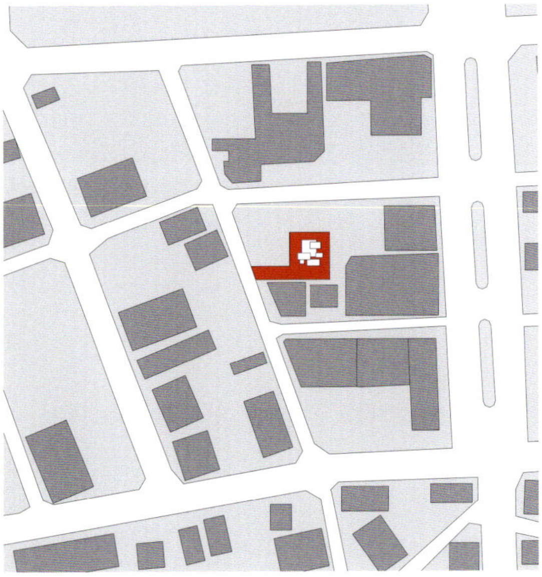

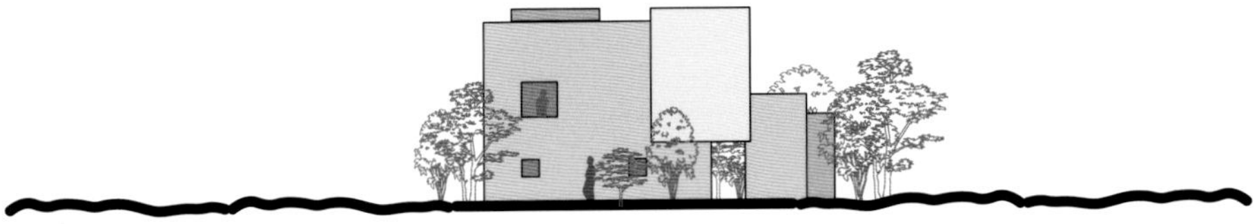

Elevations

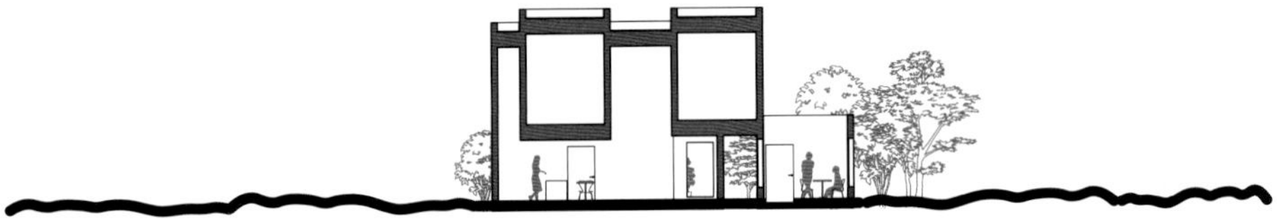

Section

This is a small residential house in Akita Prefecture. The site is in the center of Akita City where urban functions are situated. Despite its urban location, this is a special place as it is surrounded by car parks. Considering that space is exposed to public eyes in all four directions, it is essential to sustain the elements of privacy, but at the same time, our aim is to drop frontality from its façade.

In order to keep privacy, it is sensible to keep the rooms within the enclosure, and it also helps with the indoor environment in the northern climate. However, the enclosure would give a closed feeling, hence the task is to find a way to layout spaces that are closed enough to keep privacy and would also at the same time give an extensive feeling to space outside.

First, we made a list of facilities that would require enclosure; kitchen, bathroom, toilet, bedroom and storage room, all of which would be enclosed in box-shaped spaces. A collection of these spaces are gathered like pleats, which create many gap areas in various sizes. This complex structure of these gap spaces gives depth to the whole place as well as an illusion that there is more space beyond what is visible. The frames installed in the walls of the boxes overlap with one another, and the whole place looks like a combination of facing mirrors depicting different sceneries.

We designed this house on simple rules based on 2 factors: rooms that need to be enclosed should be kept in boxes and rooms that do not need to be enclosed are in the gaps between boxes. By applying these rules, space is presented with a complex structure that gives a feeling of extensive space.

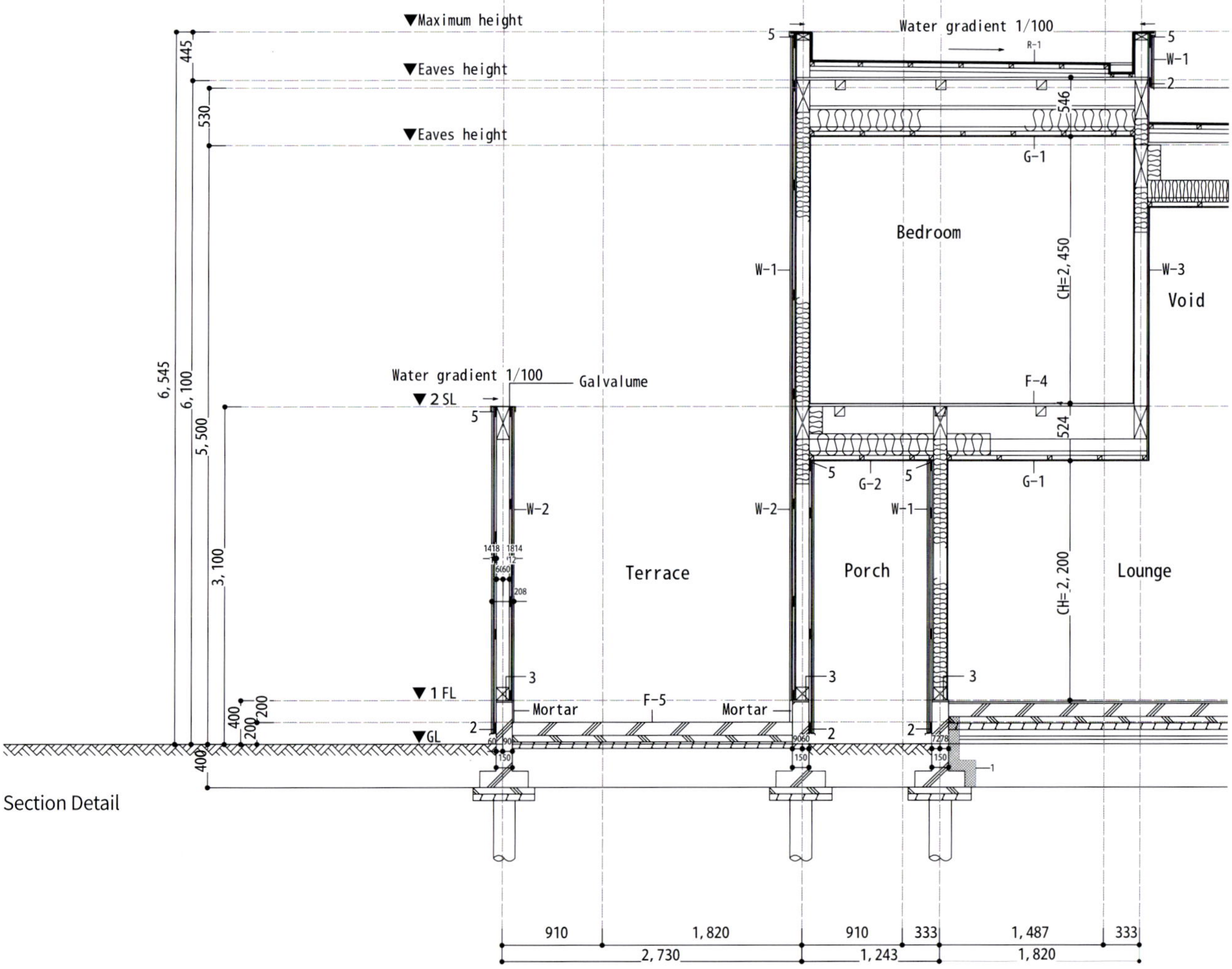

Section Detail

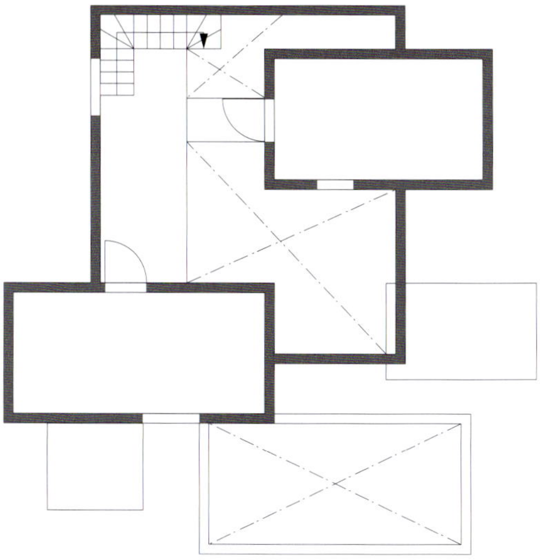

First Floor

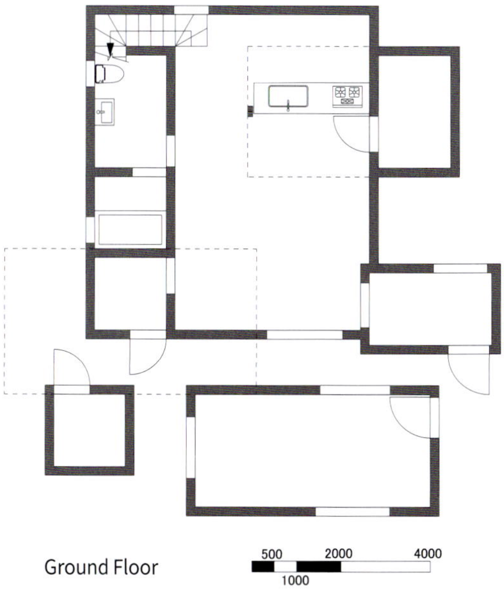

Ground Floor

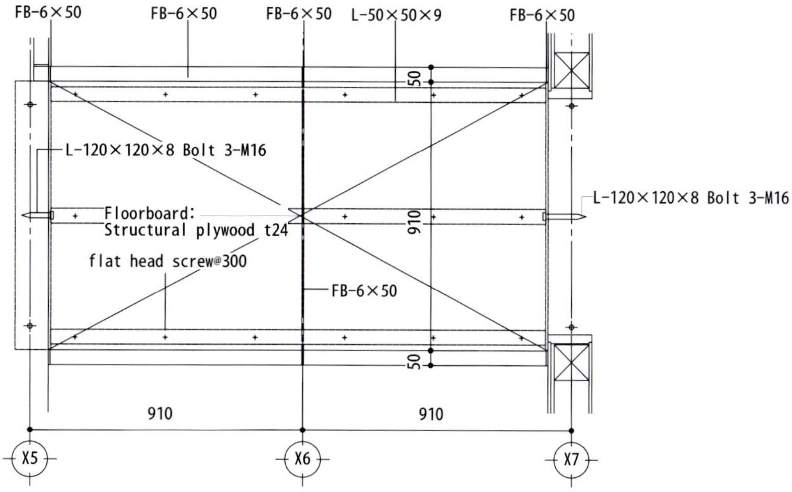

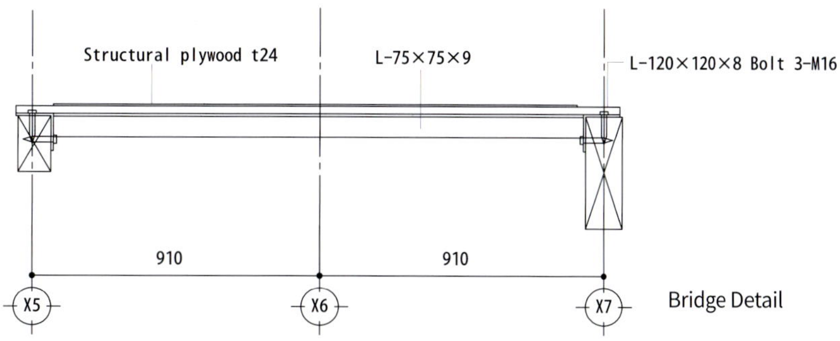

Bridge Detail

| **101.56**sq.m. | **Timber** |

COVER
by APOLLO Architects & Associates

Location _ Fuchu city, Tokyo
Site Area_ 190.01m²
Building Area _ 57.01m²
Total Floor Area _ 101.56m²
Structure _ Timber
Scale _ 2 storey
Structure engineer _ Masaki Structure, Kenta Masaki
Lighting engineer _ Sirius lighting office, Hirohito Totsune
Kitchen design_ SieMatic
Interior coordination_ Cassina ixc.
Constructor_ Hirano Construction
Photography _ Masao Nishikawa

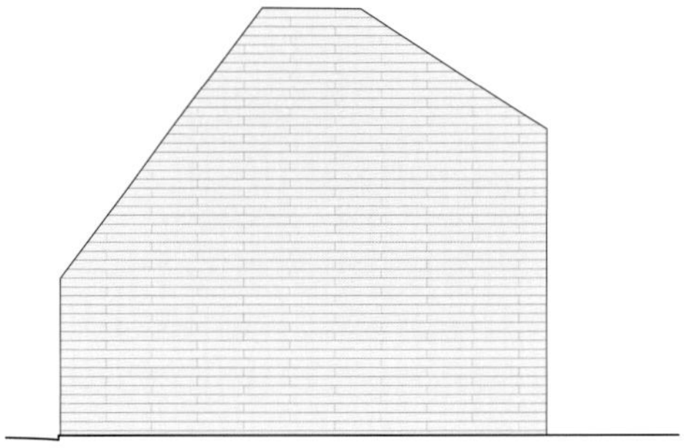

East Elevation

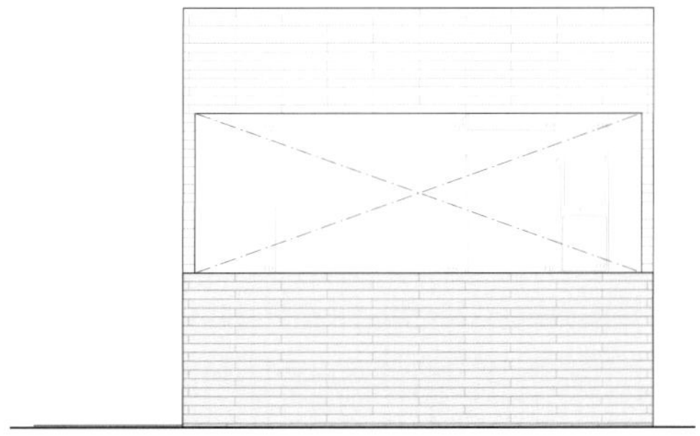

South Elevation

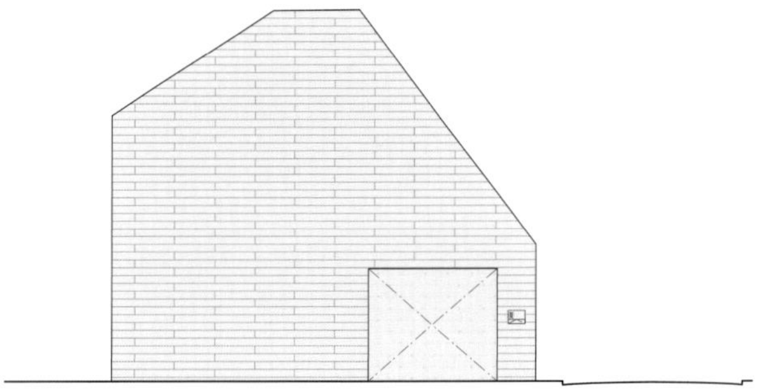

West Elevation

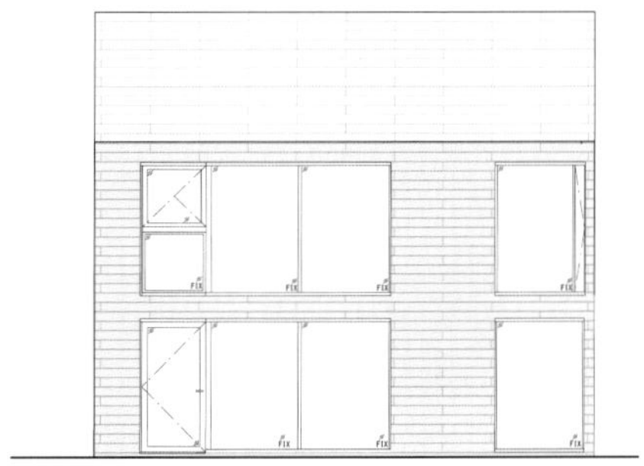

North Elevation

This home in a high-density residential district in Tokyo's suburbs was commissioned by a designer at a large real estate company. The roof and exterior walls are constructed from metallic grey Galvalume which both forms a structural skin and protects the privacy of the residents by shielding the first floor from visibility from the street. A large opening in the roof brings in abundant natural light and makes the compact space feel more expansive.

The first-floor entrance offers a view between the staircase and the bathroom clear to the back of the property. The "public dining room," an extension of the entryway, serves not only as of the family eating area but also as a guest room for entertaining. An island kitchen by Germany's SieMatic brand blends form and function, evoking a bistro atmosphere.

Through a glass door and floor-to-ceiling windows, a terrace equipped with an outdoor sofa and other features serves as an exterior living room perfect for parties. Thanks to a courtyard-like layout that ensures privacy and a peaceful mood, it's hard to believe the street is just on the other side of the wall.

The second floor is integrated with a loft and large void, forming a living area that takes advantage of the high ceilings. The interconnectivity of the exposed rafters and eaves opens up the space, while at night, soft indirect lighting envelopes it in a dream-like atmosphere. Overall the house is like an auberge where family and friends can enjoy a meal at the "restaurant" below and then relax upstairs.

BEDROOM
LIVING ROOM
ENTRANCE HALL
BATH ROOM
DINING KITCHEN

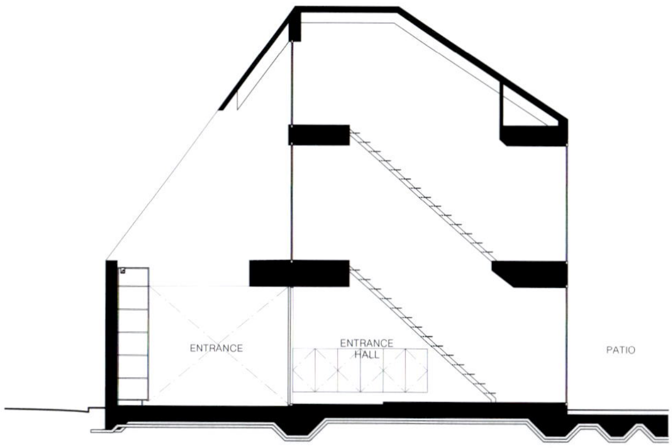

ENTRANCE
ENTRANCE HALL
PATIO

Sections

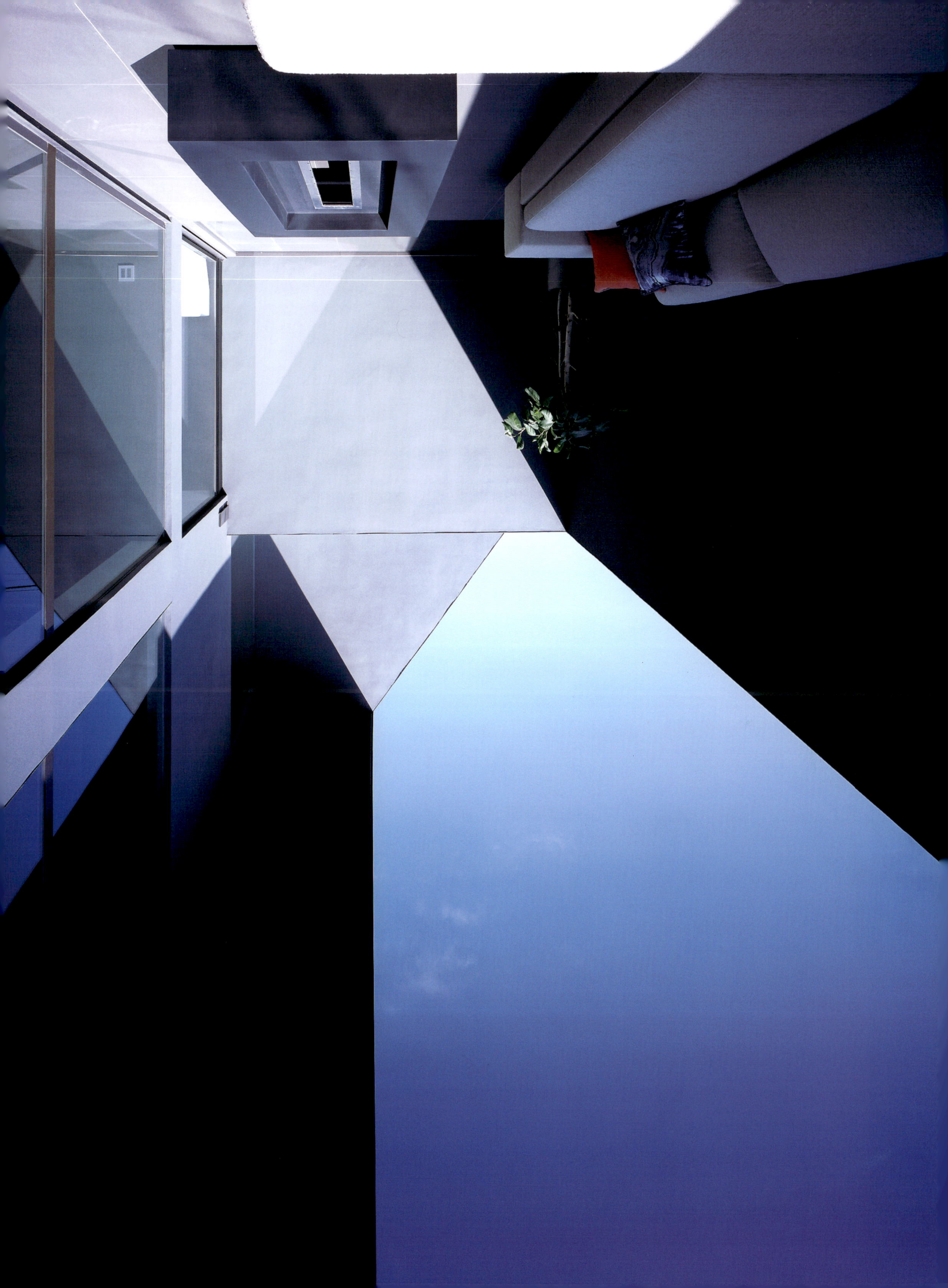

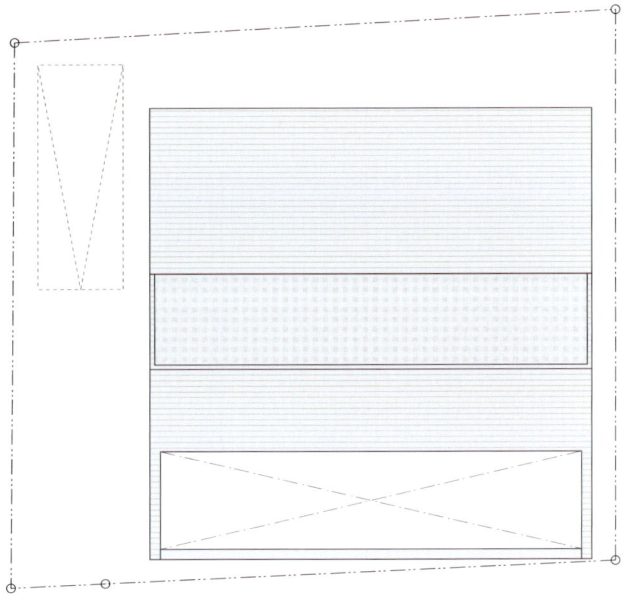

Roof Floor

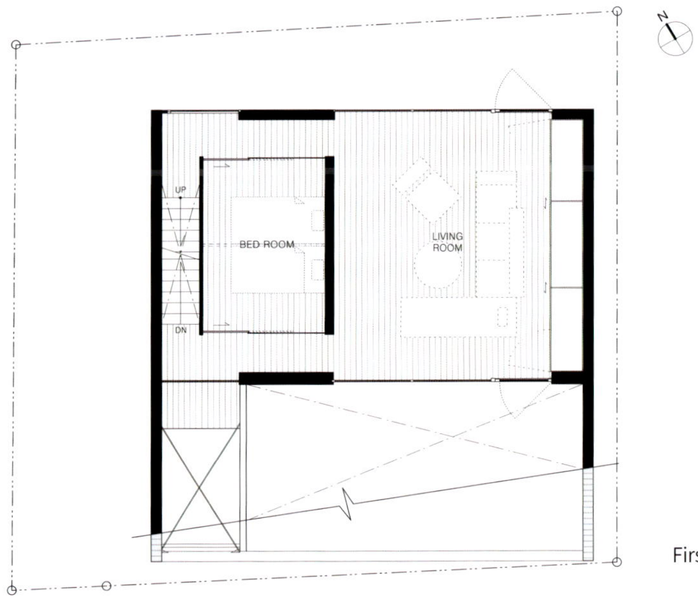

First Floor

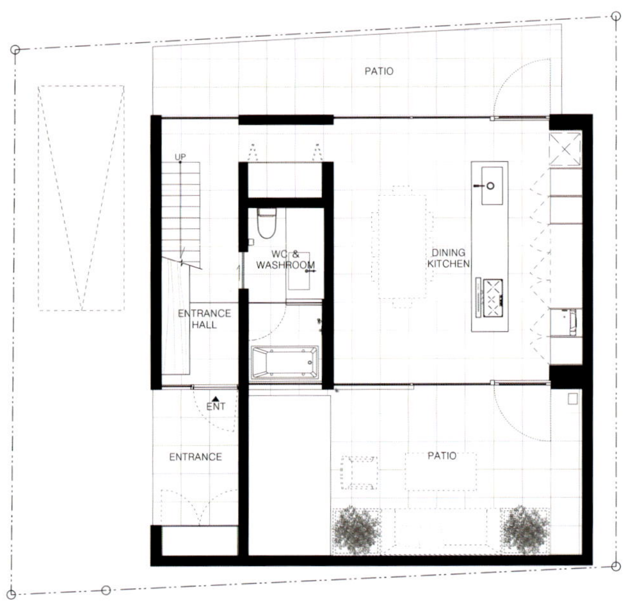

Ground Floor

| **104**sq.m. | **Timber** |

Carlton House
by Tom Robertson Architects

Location _ Melbourne, Australia
Total Floor Area _ 104m²
Styling _ Bek Sheppard
Photography _ Derek Swalwell

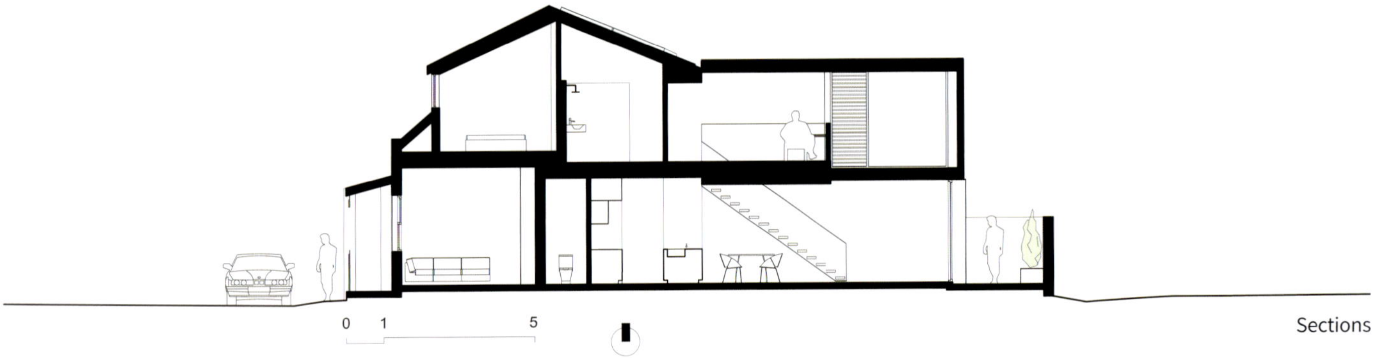

Sections

The Carlton House is a refined, light-filled renovation and extension of a Victorian terrace house. The key challenge was bringing light and space to the small, dark rooms of the original terrace. A bold roofline floods the interior with natural light from above. The design makes use of vertical space through a double-height void, which connects the living areas to the floating study. By expanding the space vertically, the home becomes open and spacious while maintaining the definition between rooms and areas of the house. A minimalist palette of crisp white, natural timber and black create a sense of simplicity and calm. Textures and decorative elements, such as the white wire mesh serving as the stair balustrade, are simultaneously functional, adding detail without over-complicating the interiors.

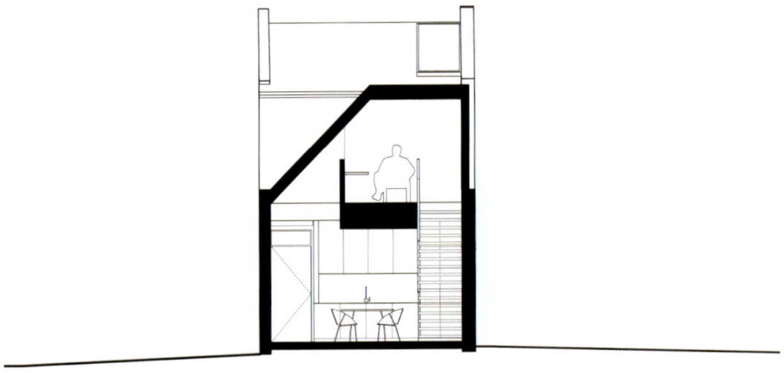

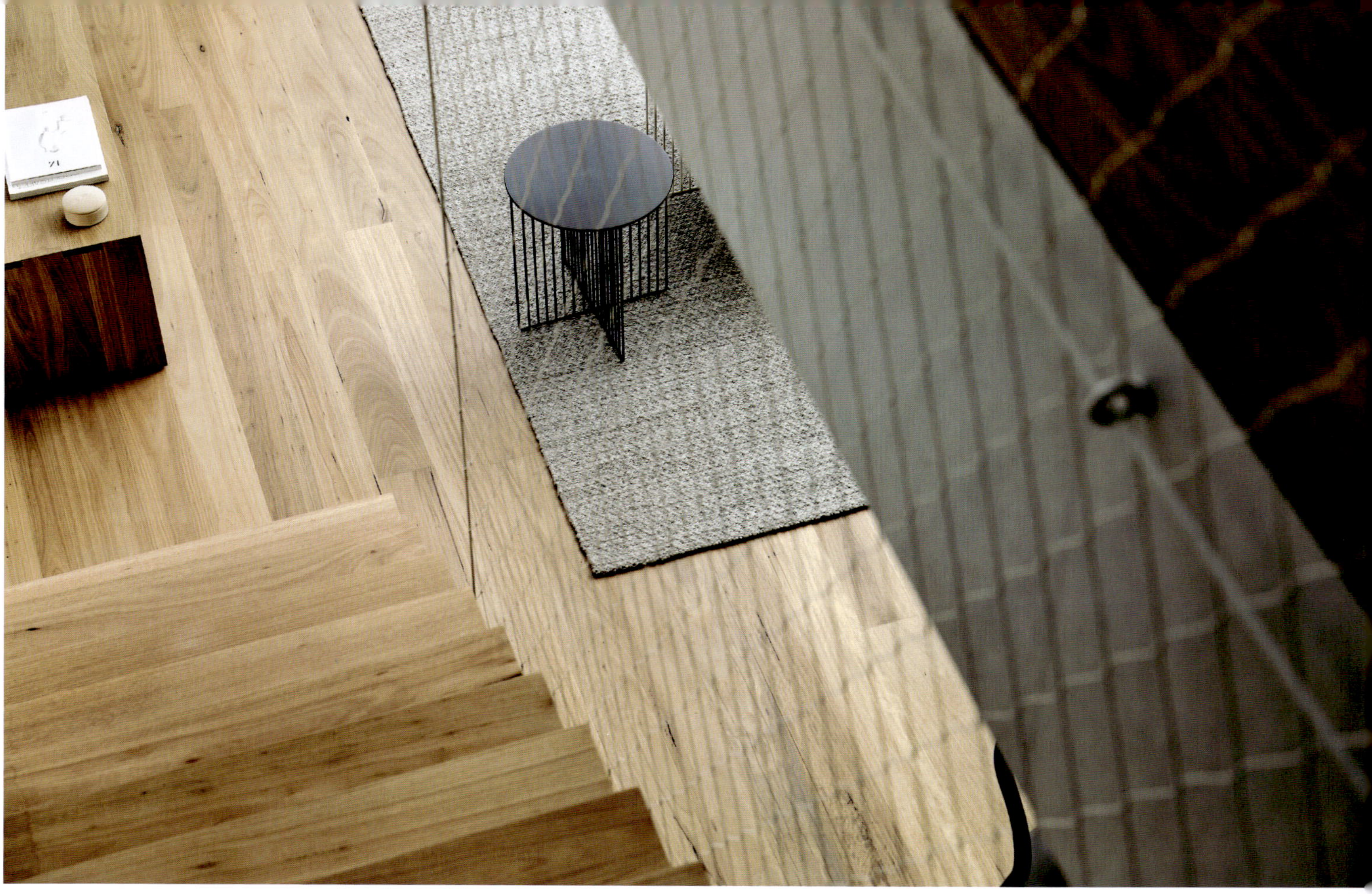

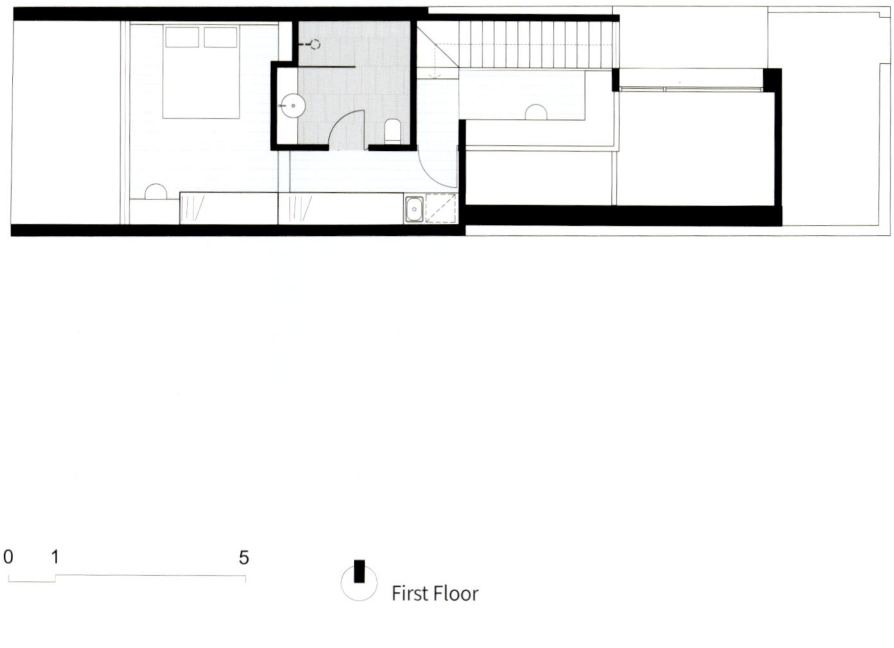

0 1 5
First Floor

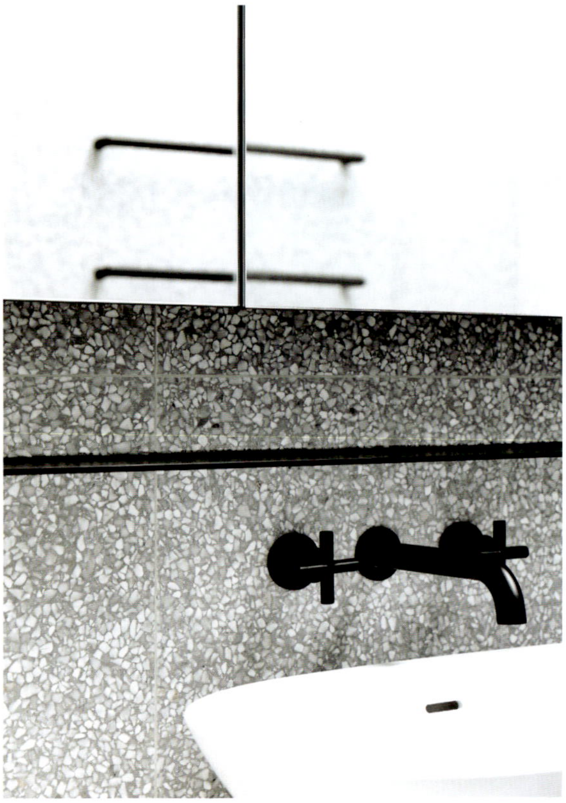

| **105.57**sq.m. | **Wood** |

Otsu House
by ALTS Design Office

Location _ Ogaya, Otsu, Shiga, Japan
Site Area _ 160.45m²
Building Area _ 55.89m²
Total Floor Area _ 105.57m²
Structure _ Wooden
Photography _ Masahiko Nishida

This house was built in the newly-developed subdivision surrounding new housing. While planning a design with the client many times, we hit upon a keyword: "comfy house" so we considered how to make a welcoming atmosphere with lots of light. With a vault -shaped opening in the simply laid-out compartment, the space is designed to be airy and rhythmical. The light spilling through a large curved window will encompass the room in a comfy manner. You can enjoy the light and shadows made by sunrise and sunset. This "comfy house will lead you to an unusual world.

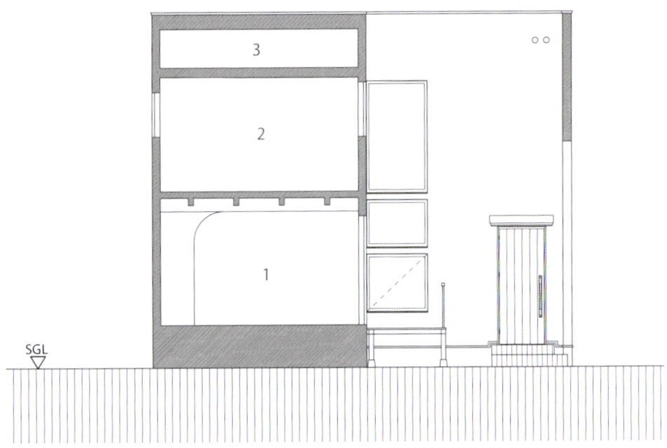

Section

1 living 2 children's room 3 loft

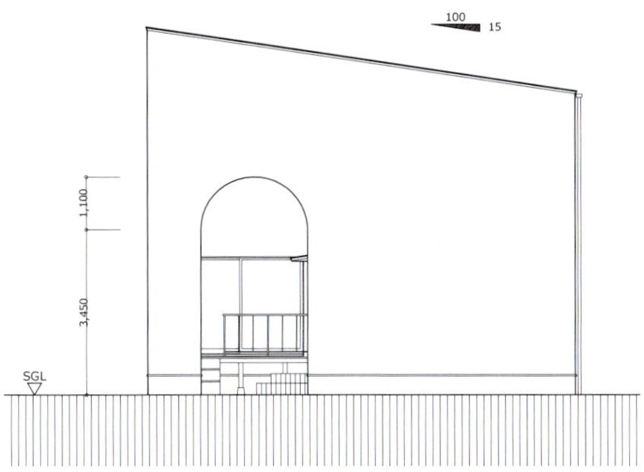

South Elevation

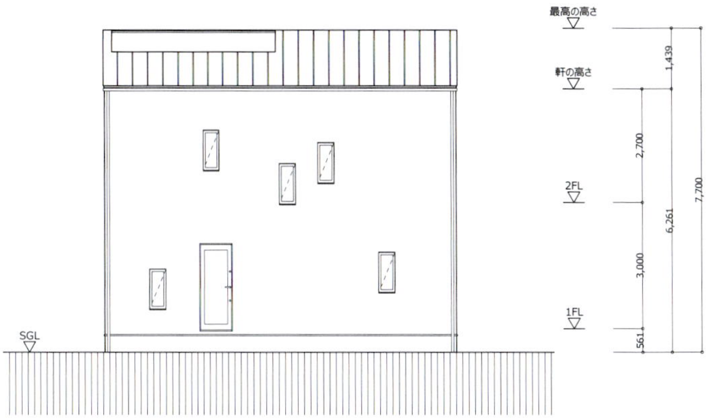

East Elevation

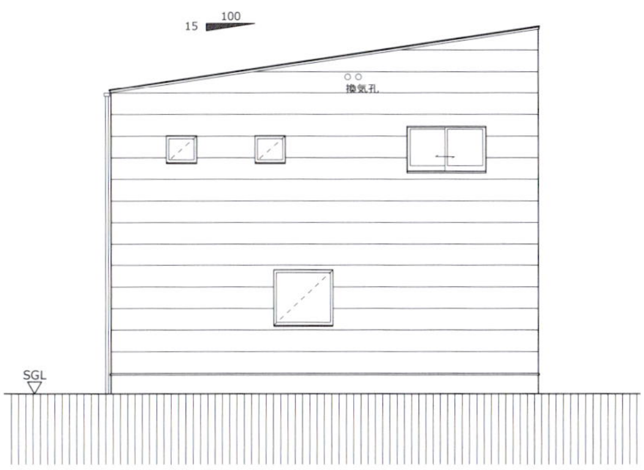

North Elevation

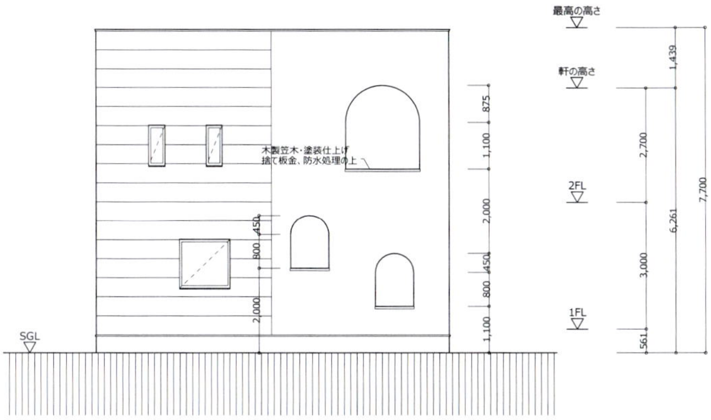

West Elevation

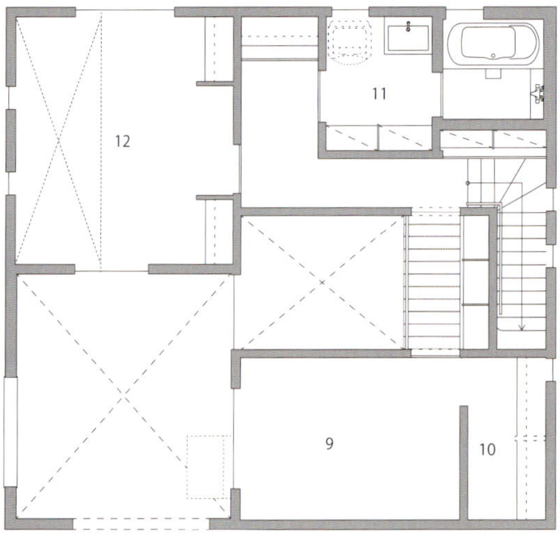

First Floor

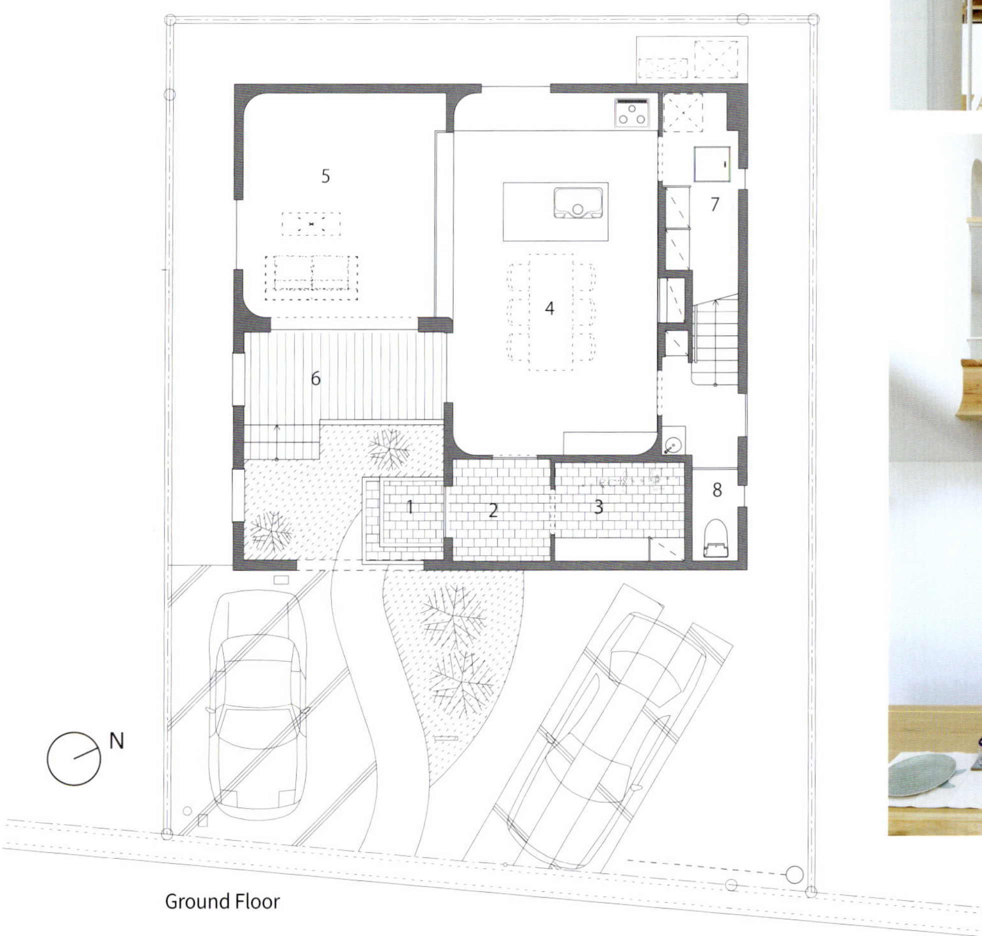

Ground Floor

1 pouch 2 entrance 3 entrance storage 4 dining kitchen
5 living 6 wood deck 7:pantry 8 WC
9:bed room 10 WCL 11 wash room 12 children's room

| **106,4**sq.m. | **Timber** |

Adorable House

by FORM/Kouichi Kimura Architects

Location _ Tokyo,Japan
Site Area _ 97.28m^2
Constructed Area_ 106,4m^2
Photography _ Yoshihiro Asada

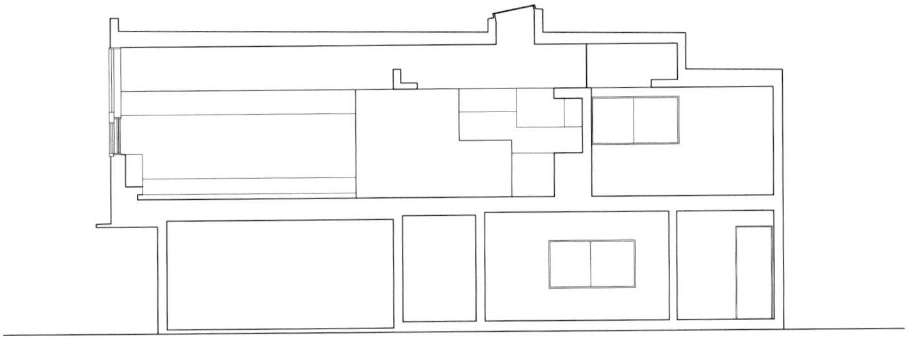

Section

The house was designed for a couple and children on 97 m2 of relatively a small ground in an urban district of a city center. The site is surrounded by newly built apartment buildings and various types of houses mixed together. Across the street, there is a park where parents can have their children play. For this project, it was requested to build a house that incorporates both functionality and design properties even if it may be small while maximizing the environment of the site. The appearance has been designed to include both massive and floaty feel that is generated by manipulating volume and proportion of configuration. The piloti space under the cantilevered eaves is used as a parking space, porch, and approach to the entrance. The space between the porch and the border of the street is used for planting to effectively display the building as well as to give a feel of linkage with the park. A small room positioned at the center of the site is used as a gallery run by the client. The room is connected to the entrance hall to secure a smooth traffic line for residents. An opening is provided on the approach side so that one can check who comes in/out or what happens outside. The internal space on the second floor that has secured a maximum height in accordance with the roof shape is produced by the light coming in from a top light and space continuity, which results from eye-conscious design. The distinctive large window facing the park is composed of a high window that introduces enough natural light, and an opening that takes in gentle light and breeze while blocking eyes from outside. The house fixture designed to be incorporated in the window frame connects the spaces, and fills the room with dignity. Scenes of children appearing at the small window facing the park resemble little birds poking their heads out of a nest, which looks so adorable. This house produces various beautiful scenes in the daily lives of the residents.

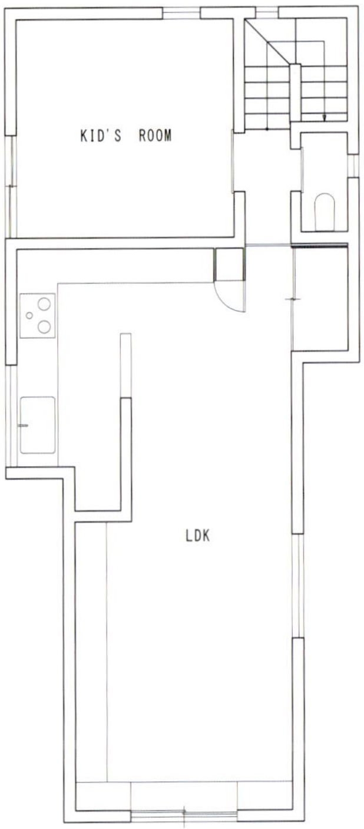

First Floor

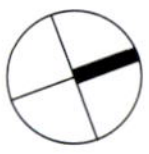

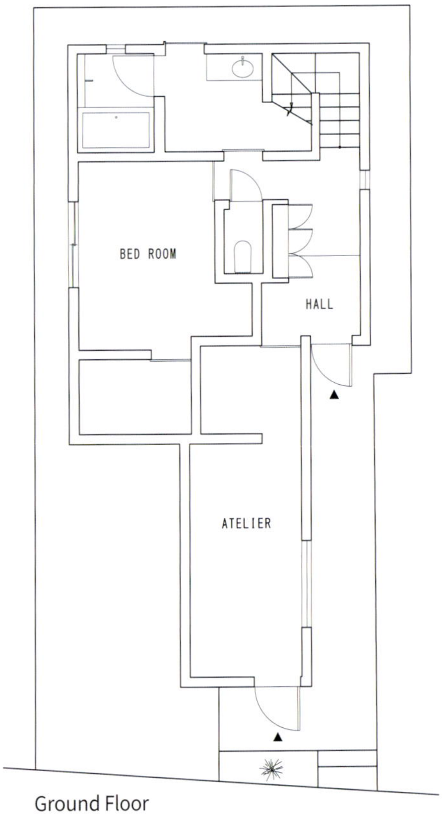

Ground Floor

| 107.62sq.m. | Concrete & Wood |

House in Toyonaka

by Fujiwaramuro Architects

Location _ Toyonaka city, Osaka Pref., Japan
Lead Architects _ Shintaro Fujiwara, Yoshio Muro
Structure _ Wooden
Site Area _ 95.60m²
Building Area _ 53.42m²
Total Floor Area _ 107.62m²
Photography _ Katsuya Taira (Studio REM)

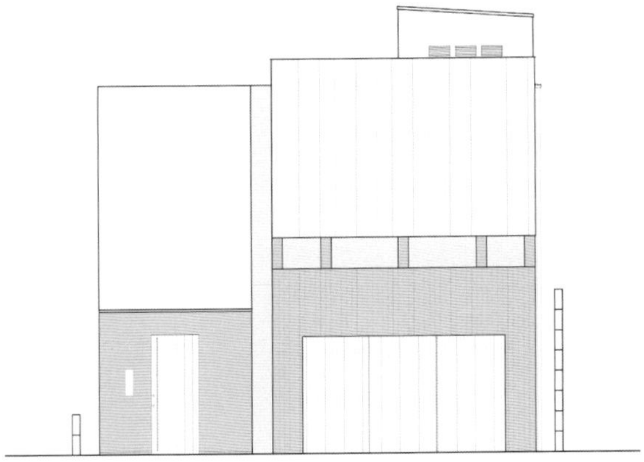

West Elevation

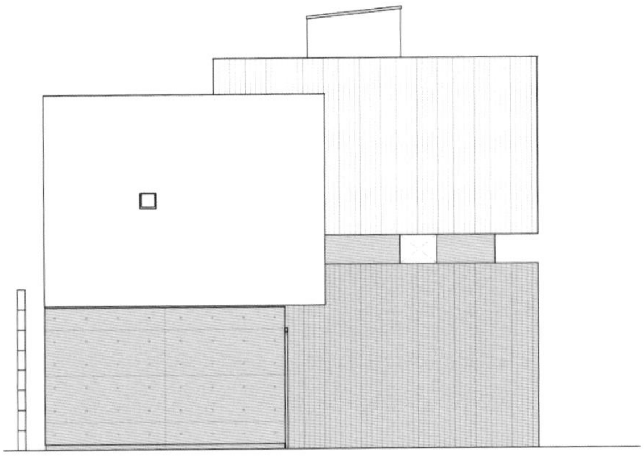

North Elevation

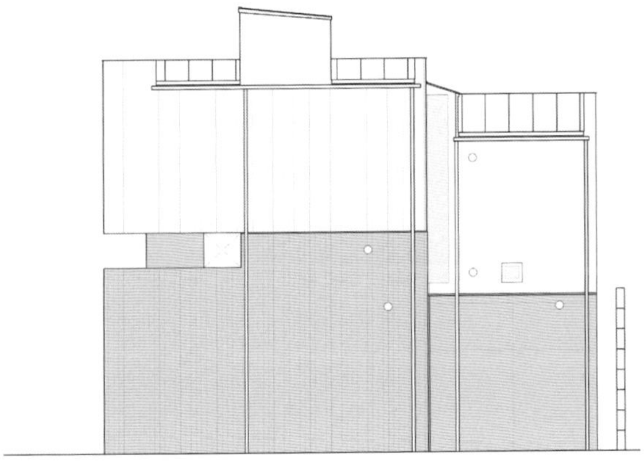

South Elevation

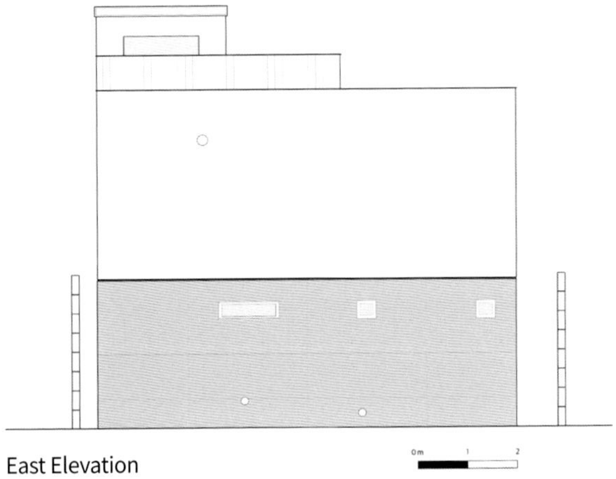

East Elevation

Located in a residential district in northern Osaka, this property is surrounded by closely packed houses and lacks any outstanding views. Due to the adjacent homes, maintaining privacy was of the utmost importance. The aim of the design was therefore to create a stimulating and beautiful space that nevertheless had no large windows and was closed to the outside world.

The building comprises a single mass divided into three boxes, the spaces between which serve as the locations for openings that provide fresh air, sunlight, and unrestricted lines of sight. The smallness of these openings is precisely what makes the sudden glimpses of sky or streetscape that they offer so striking. The way the sunlight pours in through them is also remarkable; they serve as devices for making the residents conscious that sunlight constantly shifts over time.

Outside, the use of different exterior wall materials and finishes on each of the three boxes make it easy to visually comprehend the building's structure.

The open living-dining-kitchen area is pierced by a light court that extends vertically through the building. Although the light court is constructed with flashing, outside air still passes through it, allowing it to function as an exterior space within the interior that brings light and ventilation to the main living area. Thanks to the court's glass flour, light penetrates all the way to the first-floor garage, supplementing the not-quite-adequate light coming from the gaps between the boxes. In addition, because this semi-exterior space is situated in the living-dining-kitchen area, the home avoids feeling shut-in even though it is in fact closed to the streets outside.

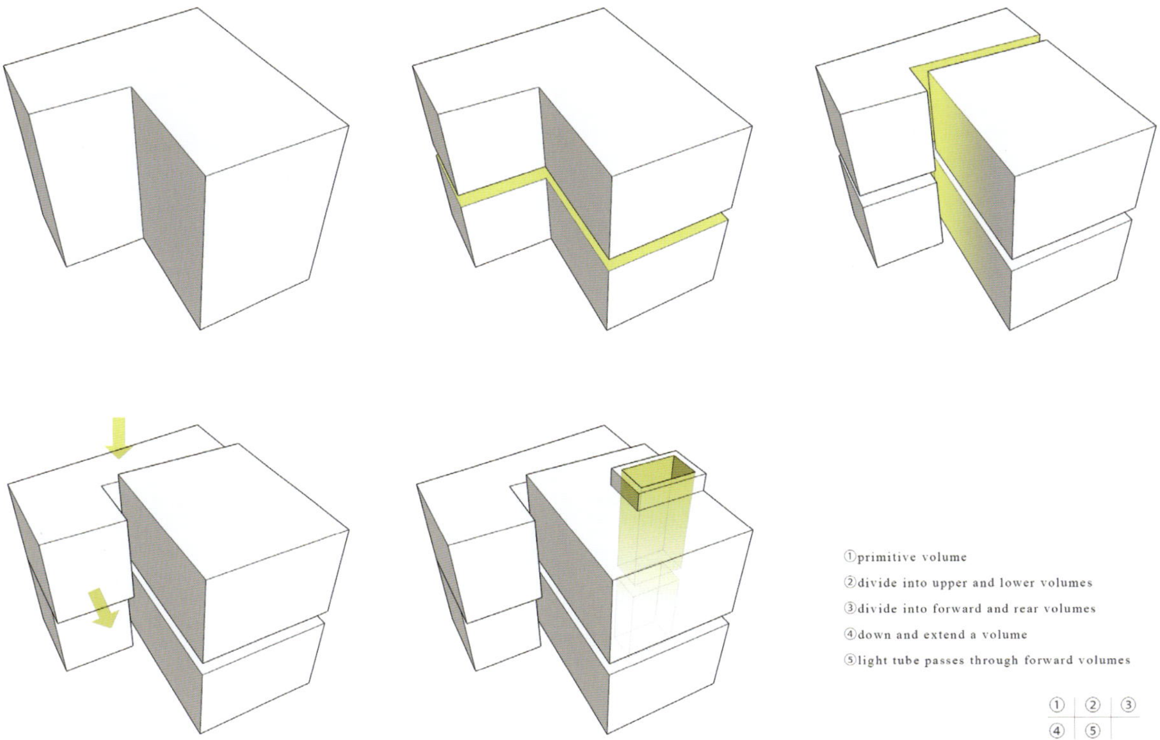

①primitive volume
②divide into upper and lower volumes
③divide into forward and rear volumes
④down and extend a volume
⑤light tube passes through forward volumes

①	②	③
④	⑤	

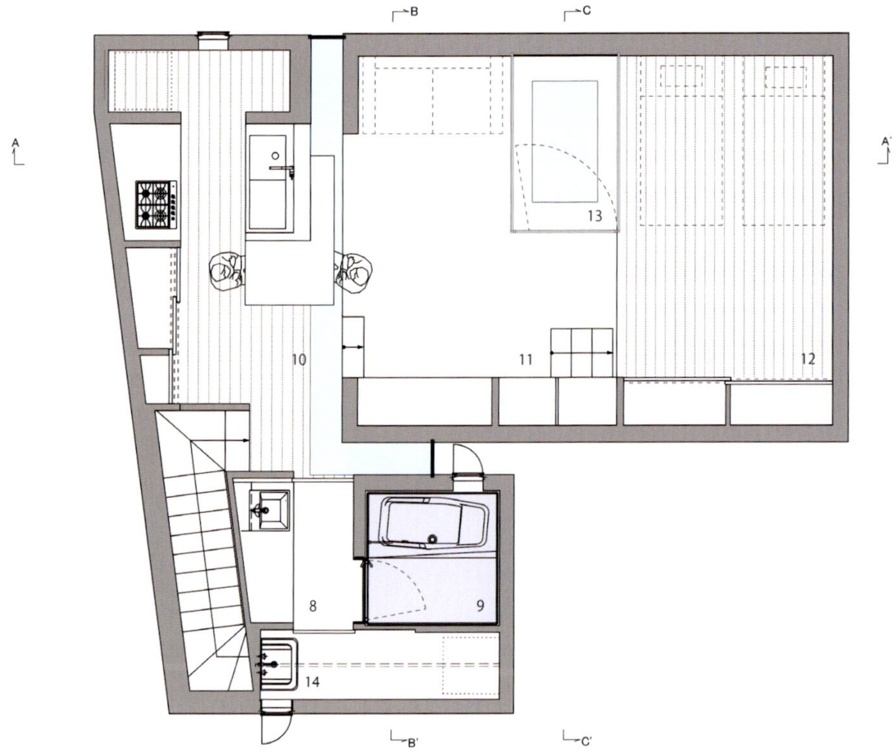

First Floor

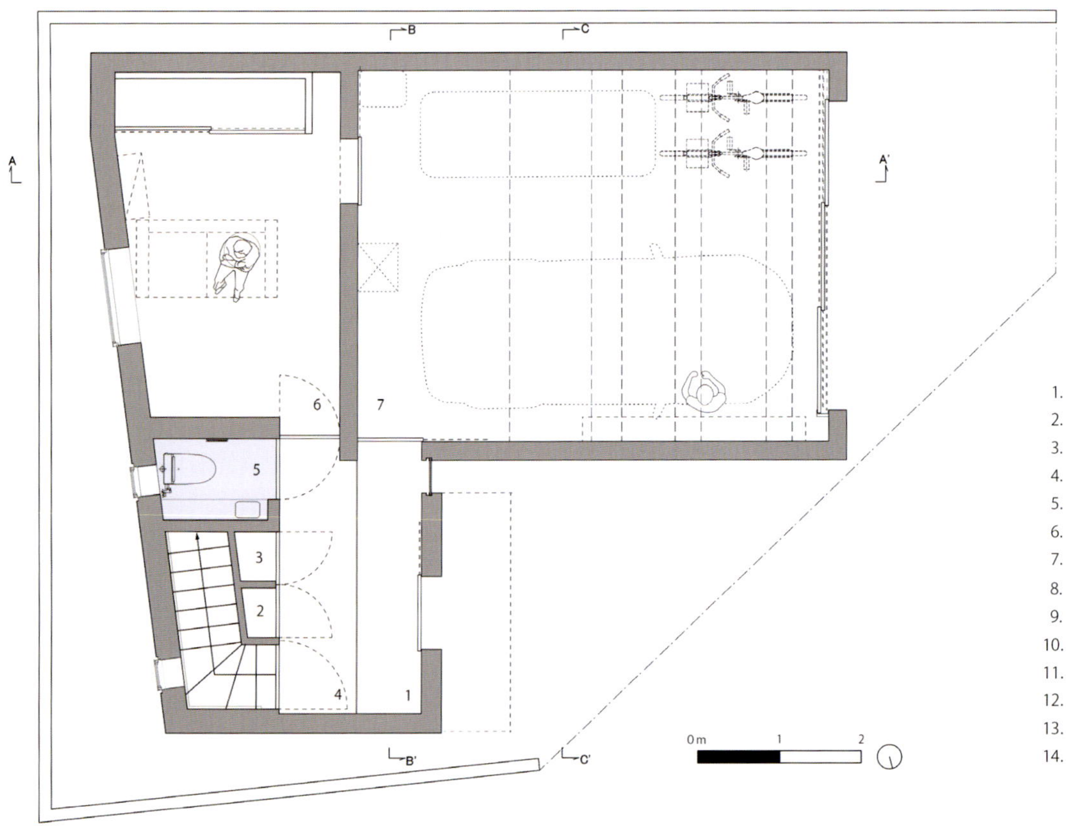

Ground Floor

1. Entrance
2. Storage 1
3. Storage 2
4. Corridor
5. WC
6. Reception Room
7. Garage
8. Sanitary Room
9. Bathroom
10. Kitchen&Dining Room
11. Living Room
12. Bed Room
13. Light Court
14. Drying Area

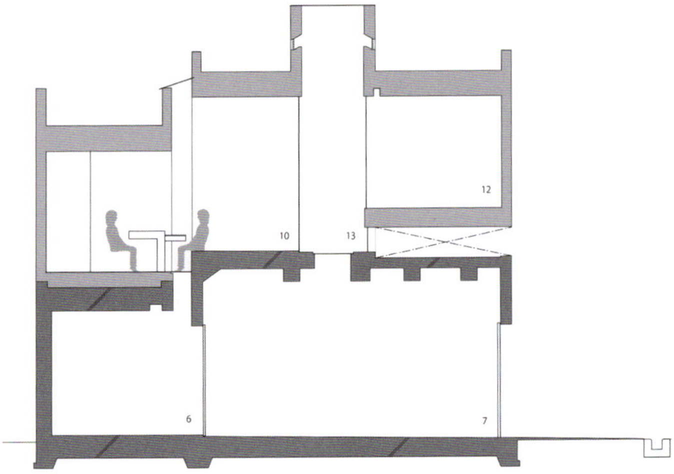

Section A-A'

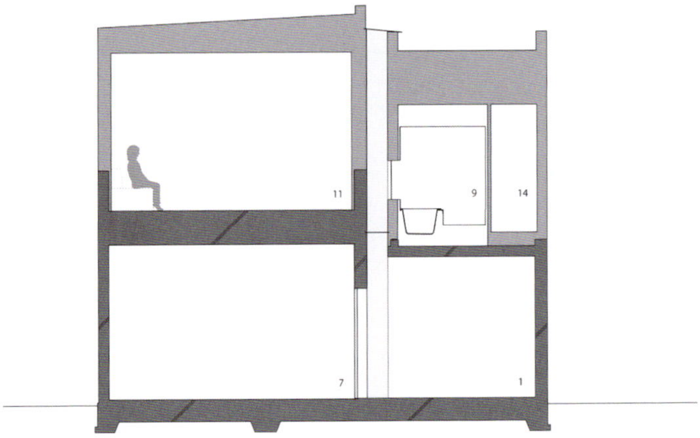

Section B-B'

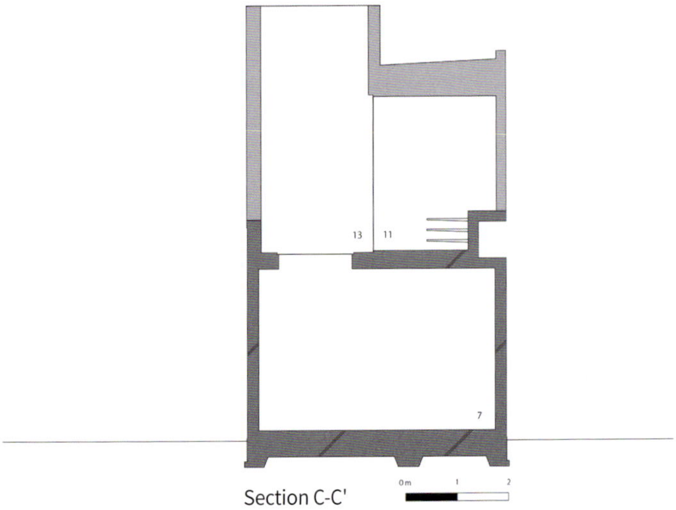

Section C-C'

1. Entrance
2. Storage 1
3. Storage 2
4. Corridor
5. WC
6. Reception Room
7. Garage
8. Sanitary Room
9. Bathroom
10. Kitchen&Dining Room
11. Living Room
12. Bed Room
13. Light Court
14. Drying Area

| **112.27**sq.m. | **Timber** |

142

House of Scenes

by FORM/Kouichi Kimura Architects

Location _ Shiga,Japan
Site Area _ 175,68m^2
Constructed Area_ 112,27m^2
Photography _ Yoshihiro Asada, Norihito Yamauchi

The site is located in a zoned residential area which is a little bit away from the bustling main street. The client wished to have a comfortable house where she could hold a private Yoga class. To meet that request, the linkage between exterior and interior spaces were identified to realize a living space that expands toward outside in terms of visual and sensory aspects. When opening the entrance door of the symmetrically designed building, a courtyard surrounding the living room stretches right in front. The living room that also serves as the Yoga classroom has two openings; it can be used as one unified space together with the courtyard when the sliding doors are wide opened. The terrace continued from the living room is used for Yoga lessons. Light reflected by the basin and fresh wind stimulate the five senses and set the mind and body free. The wall extended from the terrace incorporates a tower structure that can be regarded as the symbol of this house. The frosted glass fitted in the sculptural-shaped form makes the light thrown into the room softer. It also serves as indirect illumination for the courtyard at night. The fence surrounding the courtyard not only blocks eyes from surroundings but unites with the building to make a semi-outdoor room that can be utilized as an inner terrace. The mortar counter and bench fabricated inside extend the exterior living space and serves as an element to enhance theming effect from the living room. On the second floor, each room is laid out with a dining room at the center. Space is segmented by the difference in ceiling height. Natural light coming through the colored top light of the ceiling quietly brings an uplifting feeling. The dining room is provided with a large window surrounded by the walls and long eaves. The symbolic riser from the courtyard has been designed to serve as an eye-catching object in the scene. It also brings about visual continuity and links the living scene on the lower floor with that on the upper floor. Through the connection with exterior space in diverse ways, the house produces pleasant scenes of living.

West Elevation

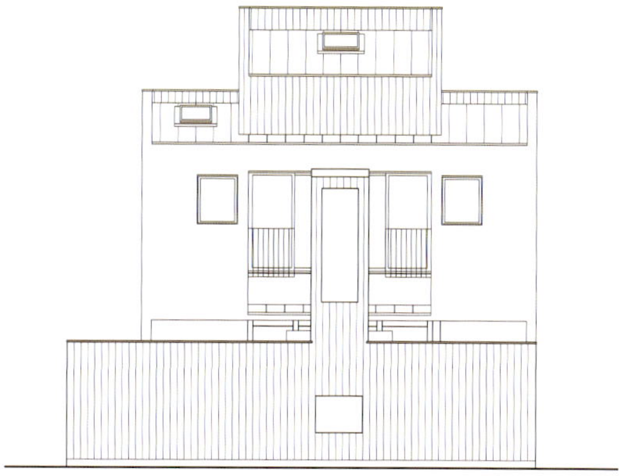

East Elevation

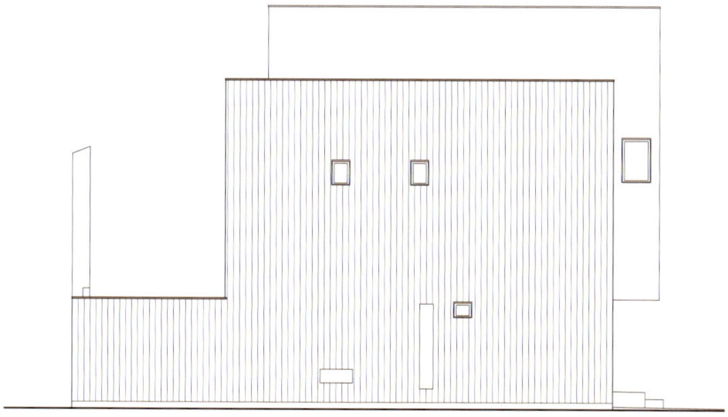

North Elevation

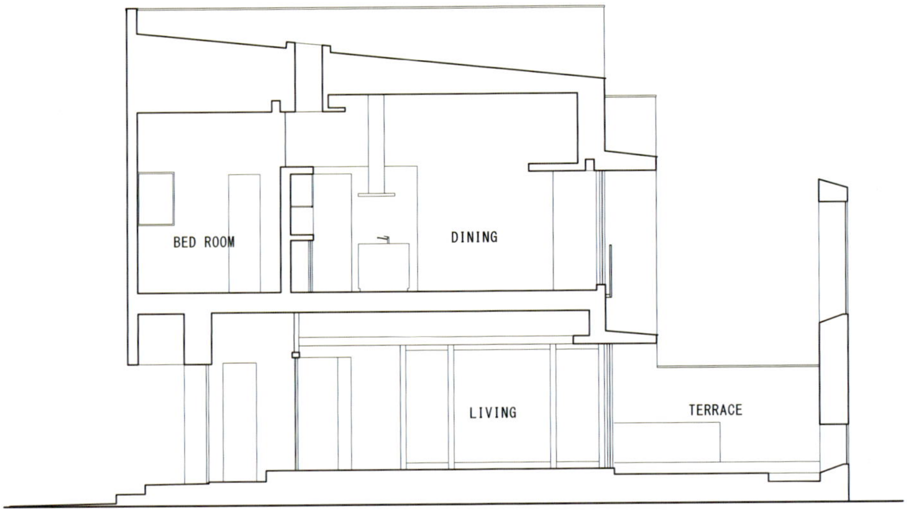

Section A-A'

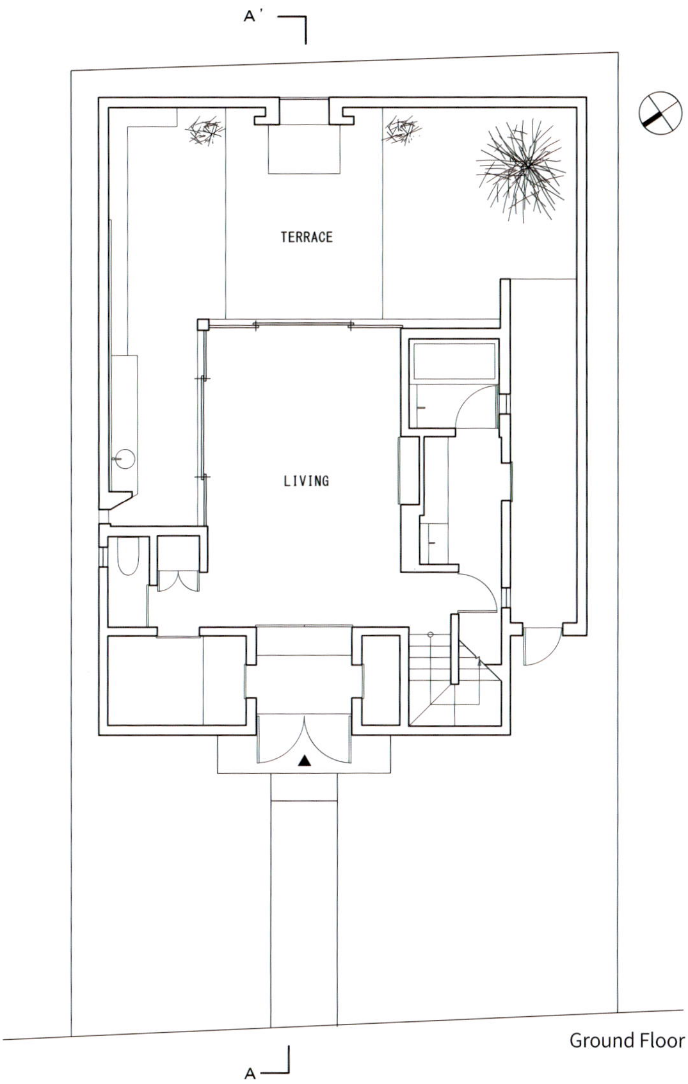

Ground Floor

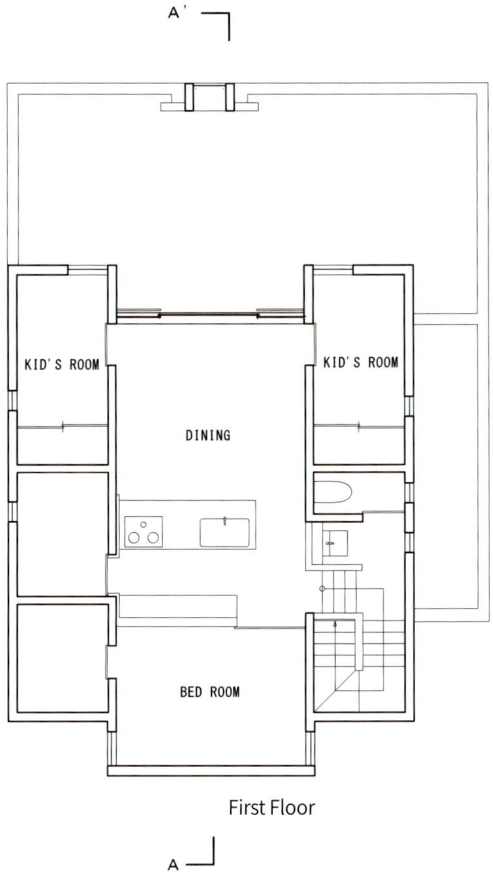

First Floor

| **112.28**sq.m. | **Timber** |

Slender House

by FORM/Kouichi Kimura Architects

Location _ Shiga,Japan
Site Area _ 125,24m^2
Constructed Area_ 112,28m^2
Photography _ Yoshihiro Asada, Norihito Yamauchi

The house is located in the subdivision accessible from a suburban arterial road. It has been planned in the strip shape lot; with narrow width and large depth. In the vicinity are the houses built by house-building companies or local builders. We then aimed to build a house that possesses dignified presence with both designability and coziness in such a cluttered townscape. The exterior is composed of volumes with carefully examined proportion, window layout, and materials. The window with height and the overhang porch emphasize vertical and horizontal lines, making the facade not too much uniform. The entrances are positioned on both sides of the porch to enhance convenience for access from the parking spaces that are laid out on either side. This moderately blocks eyes from the road and secures privacy as well. The interior of the first floor is connected from the entrances through the washroom at the penetralia, along the deep and strip-shaped lot. The raised floor, staircase, and furniture are rhythmically laid out along the passage. They are used as flexible spaces for communication to enrich daily lives. On the second floor is the space with a feel of openness, where eyes from outside are blocked on the one hand while those letting go from inside to outside on the other hand. Buildings come close in the periphery, but a ridgeline of the near mountain can be viewed to the road in the north. Making the best of this environment, a living room with high ceiling has been laid out toward the road, and a large opening provided to cut out the surrounding landscape. The natural scenery and light that are taken inside the house make the internal appearance everchanging, allowing eyes to go through and expand. In addition, the light that is reflected in the shiny wainscot makes the view in the front more beautiful. The building with impressive appearance generates cozy space inside. It is achieved by studying the potential and characteristics of the site and adding to the building the sense that is unique to architects.

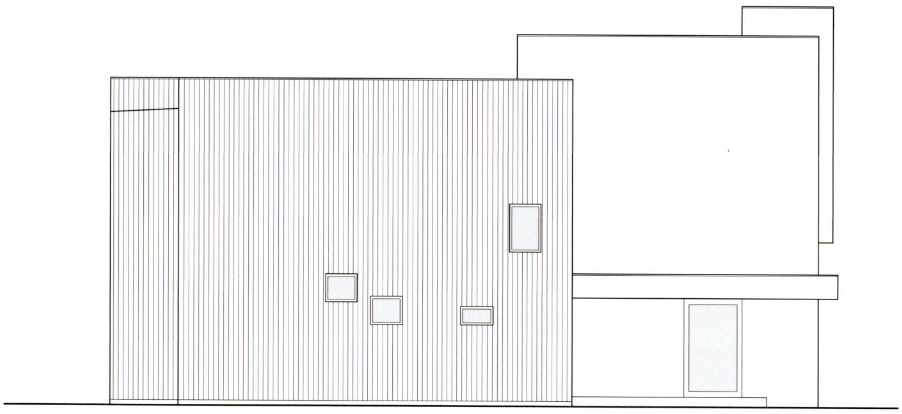

West Elevation

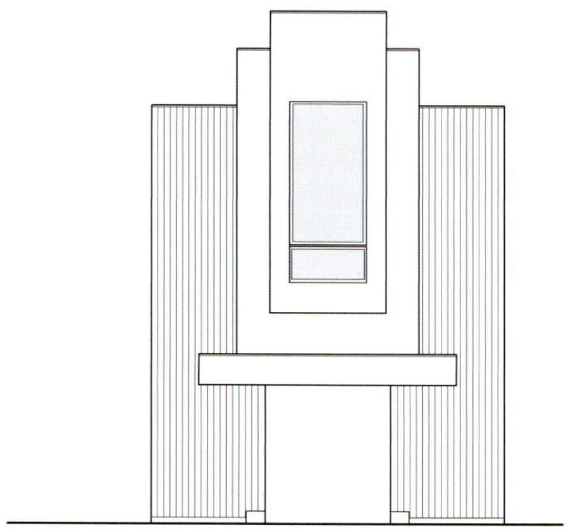

North Elevation

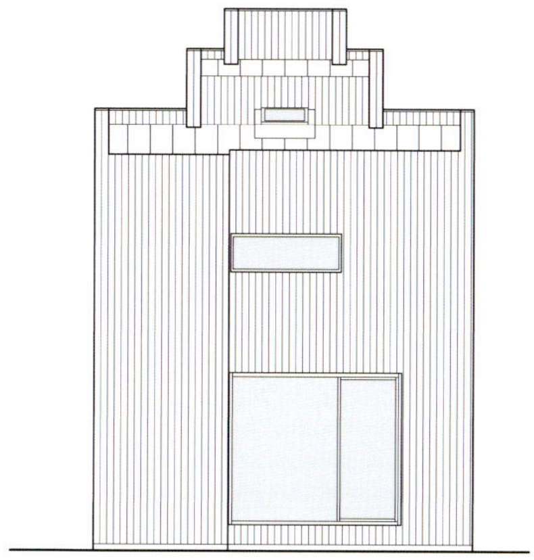

South Elevation

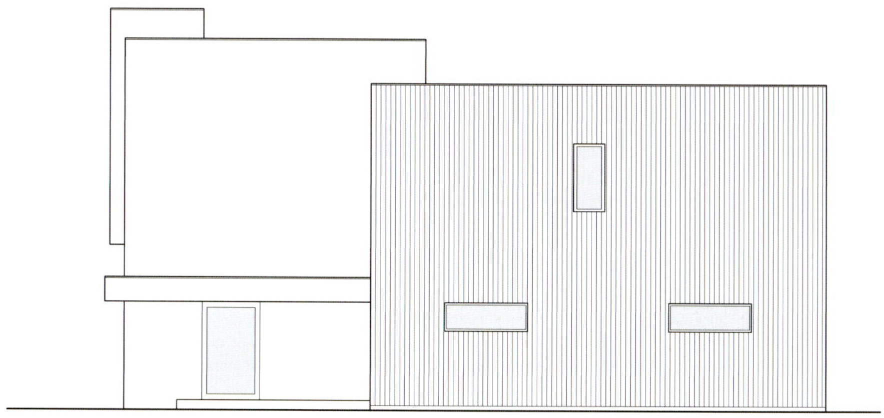

East Elevation

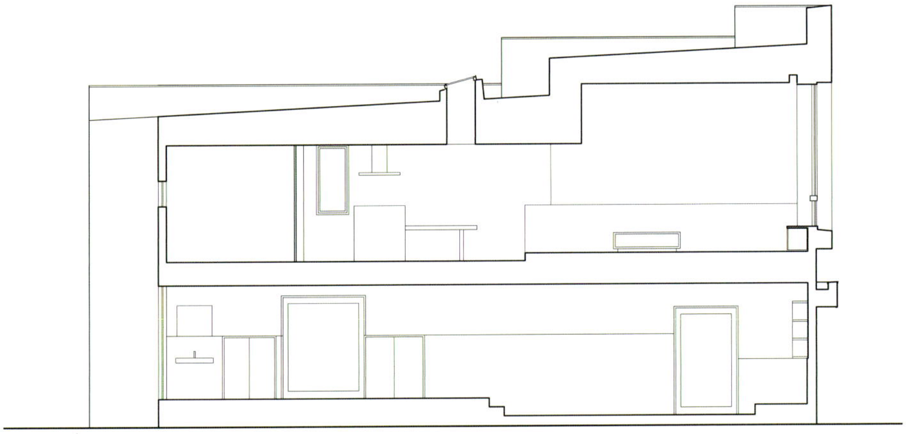

Section

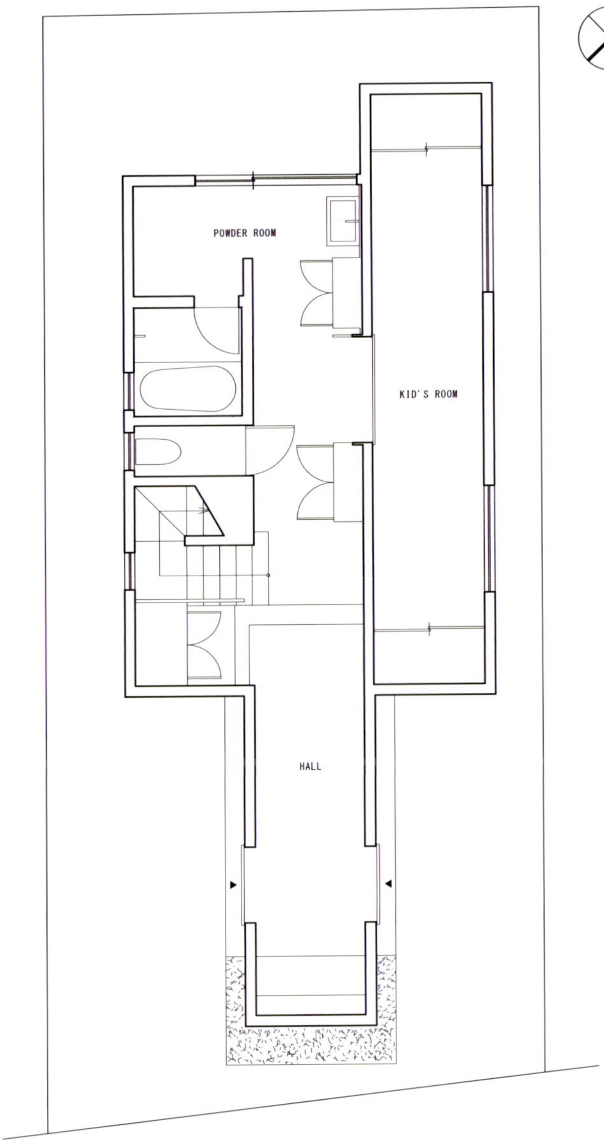

Ground Floor

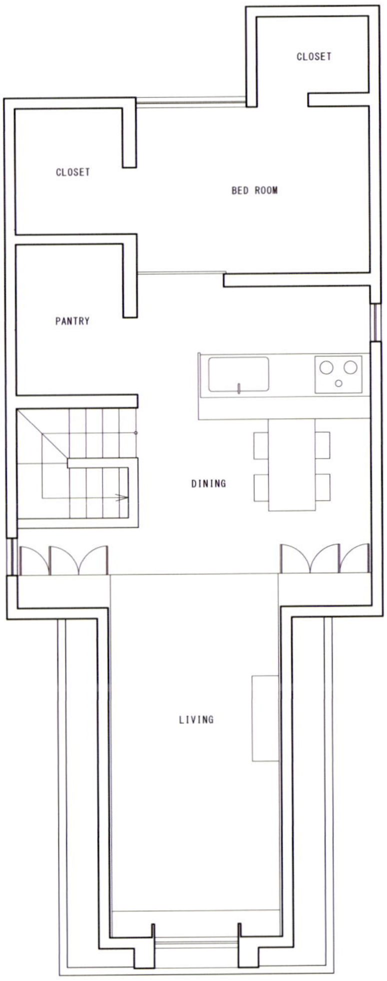

First Floor

200
3335
500
100
400
1300
950
400
400
950
150

| **116.57**sq.m. | **Timber** |

TRANE

by APOLLO Architects & Associates

Location _ Meguro ward, Tokyo
Site Area_ 72.19m²
Building Area _ 42.93m²
Total Floor Area _ 116.57m²
Structure _ Timber
Scale _ 3F+PHF
Structure engineer _ Masaki Structure, Kenta Masaki
Lighting engineer _ Sirius lighting office, Hirohito Totsune
Constructor_ Arata kenchikukoubou
Photography _ Masao Nishikawa

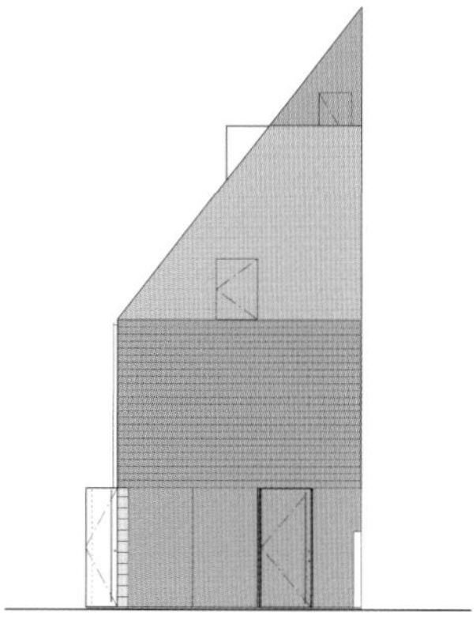

West Elevation

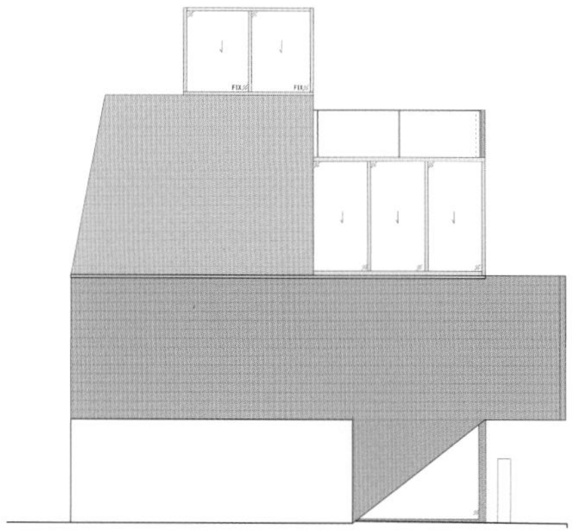

North Elevation

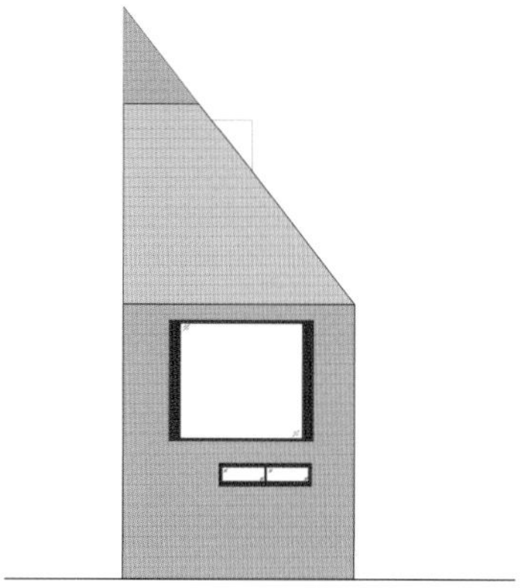

East Elevation

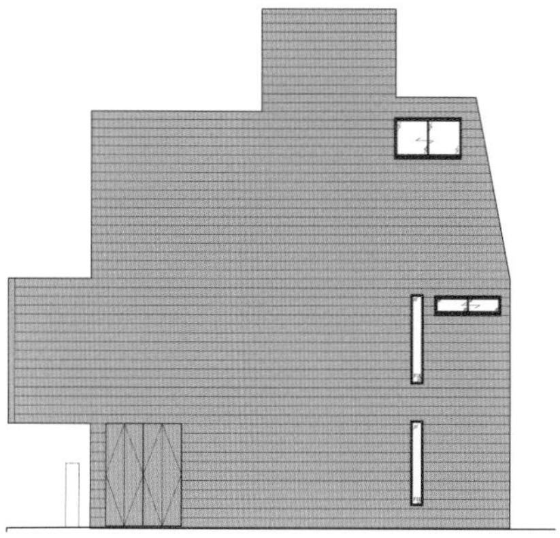

South Elevation

This small house in downtown Tokyo is owned by a husband who works in finance and his wife who works in the high-end furniture industry. The husband had previously been a jazz musician, and the starting point for the design was his request that the living space reflects the world-view of his role model John Coltrane's music.

The building is elevated by pilotis, with the entryway located next to a parking area sheltered by the overhanging volume. The master bedroom is at the back of the ground floor, with a bathroom in front filled with peaceful light that pours in from an adjacent courtyard. A staircase leads to the second floor, where a jazz-bar-like dining and kitchen area offers an ideal spot for sharing a meal or drinks with family and close friends.

The living room is defined by its slanted double-height ceiling and skylights that bring in plenty of natural light, while a catwalk bridge lends an industrial ambiance. The residents enjoy listening to jazz in this space, where a photograph of John Coltrane hangs on one wall. On a small balcony facing the road, steel louvers block visibility from below to ensure privacy.

The children's room and other private spaces are located in an open arrangement on the third floor, which also provides access to a rooftop bar where the residents enjoy garden parties with their frequent guests.

By avoiding partitions in the compact structure, the design succeeds in creating a single open room linked vertically from floor to floor. The result is a theatre-like small home whose residents live their own story.

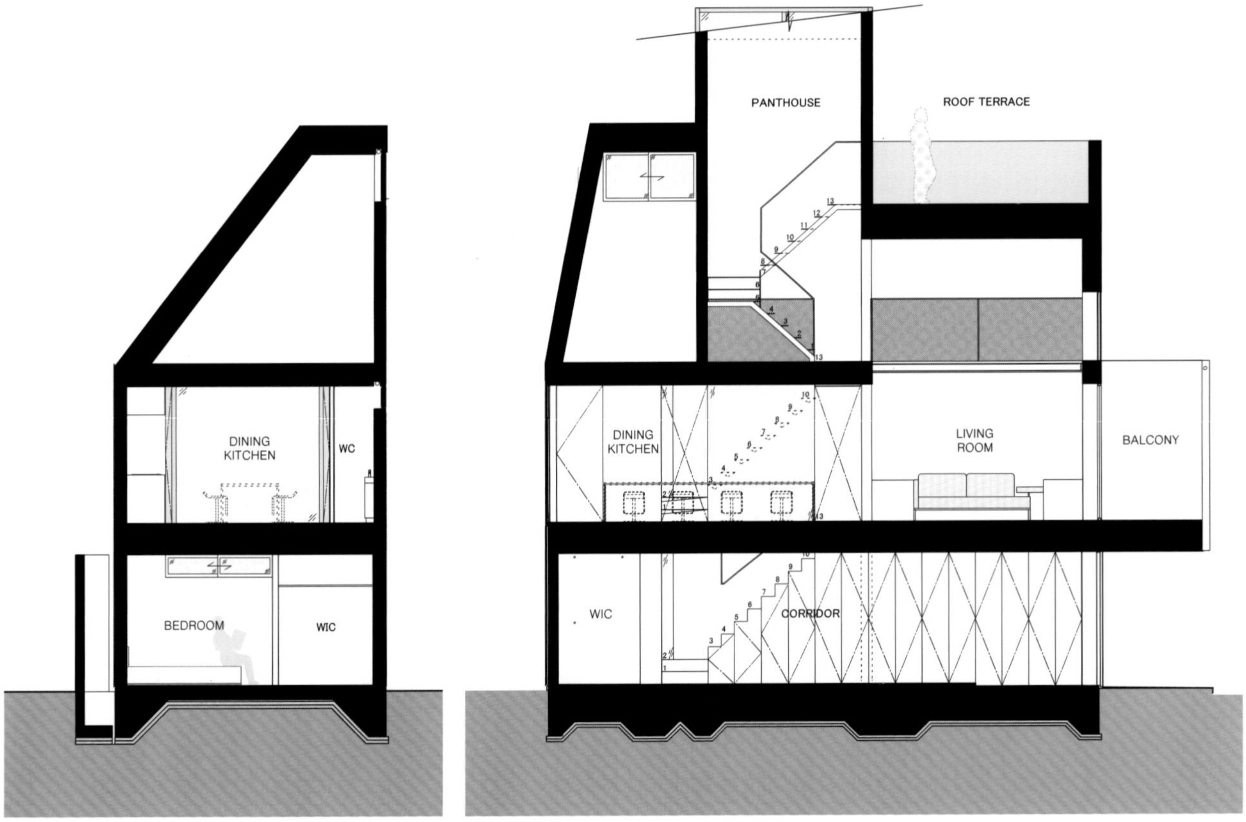
DINING
KITCHEN
WC
BEDROOM
WIC
PANTHOUSE
ROOF TERRACE
DINING
KITCHEN
LIVING
ROOM
BALCONY
WIC
CORRIDOR

Sections

BLUE TRAIN john coltrane
JOHN COLTRANE

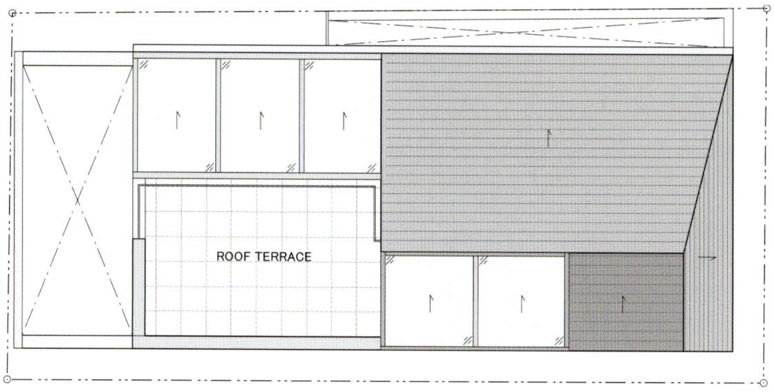

Roof Floor

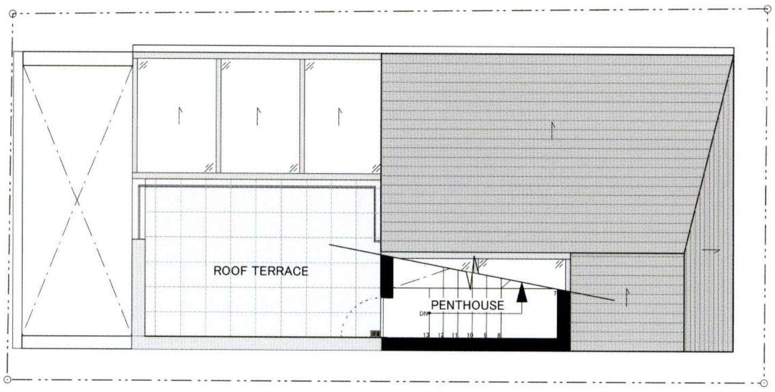

Panthouse Floor

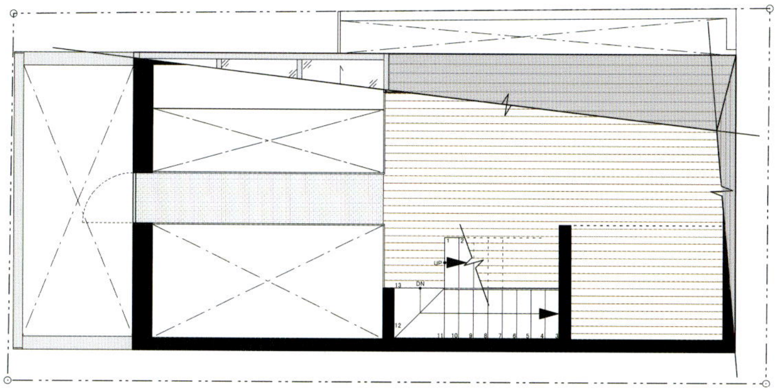
Second Floor

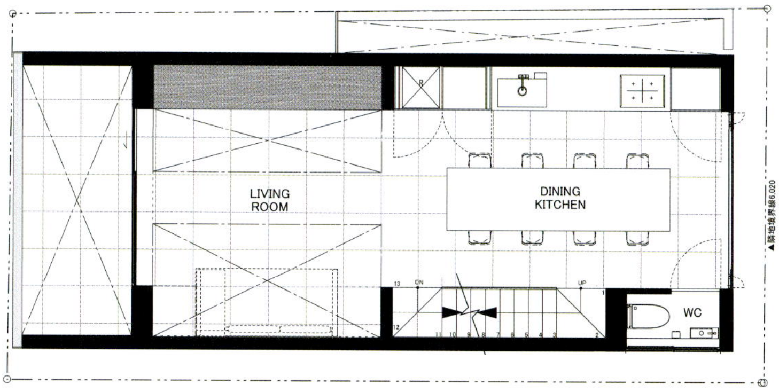

First Floor

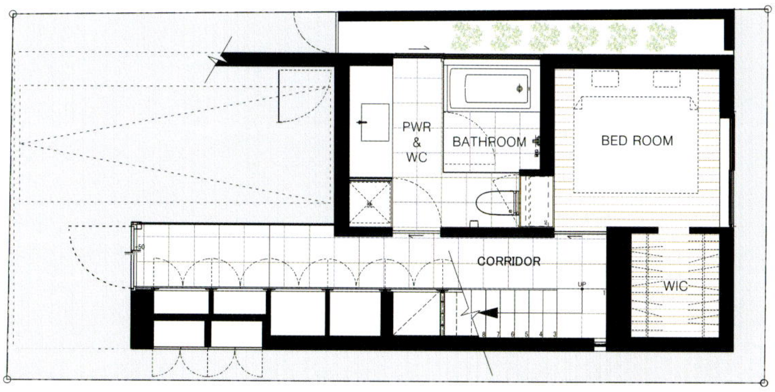

Ground Floor

BLUE TRAIN
john col
JOHN COLTRAN

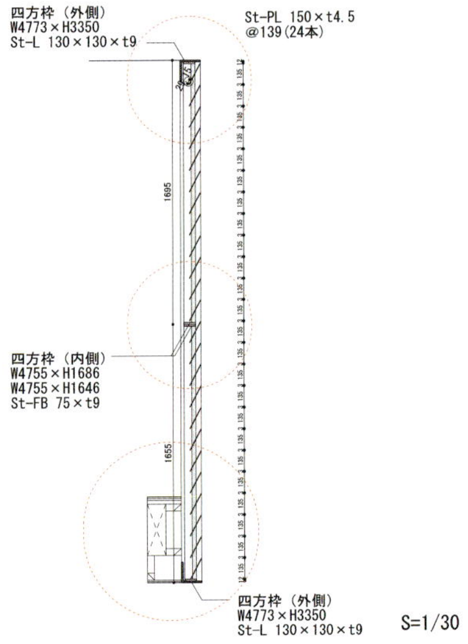
四方枠（外側）
W4773×H3350
St-L 130×130×t9
St-PL 150×t4.5
@139(24本)
四方枠（内側）
W4755×H1686
W4755×H1646
St-FB 75×t9
四方枠（外側）
W4773×H3350
St-L 130×130×t9
S=1/30

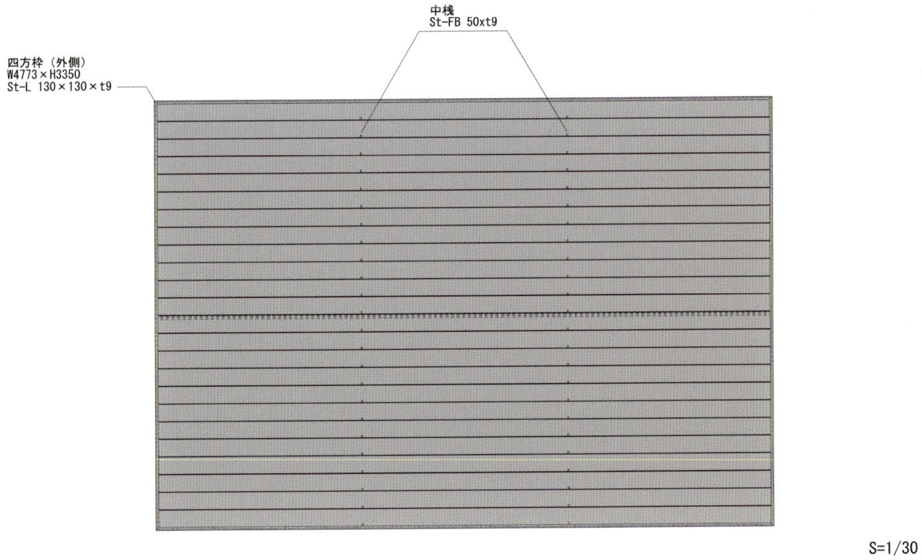
中桟
St-FB 50xt9
四方枠（外側）
W4773×H3350
St-L 130×130×t9
S=1/30

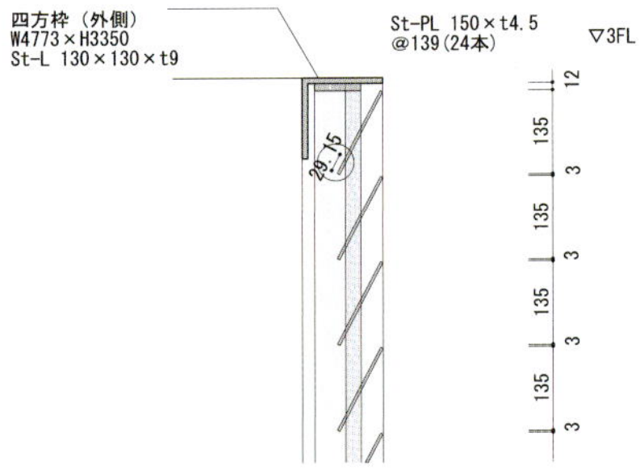
四方枠（外側）
W4773×H3350
St-L 130×130×t9
St-PL 150×t4.5
@139(24本)
▽3FL

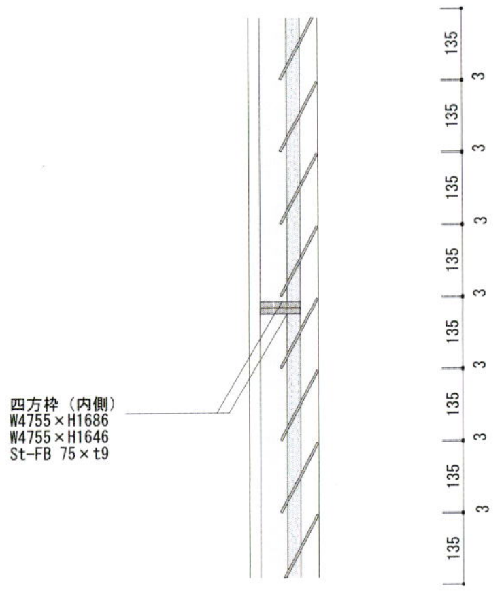
四方枠（内側）
W4755×H1686
W4755×H1646
St-FB 75×t9

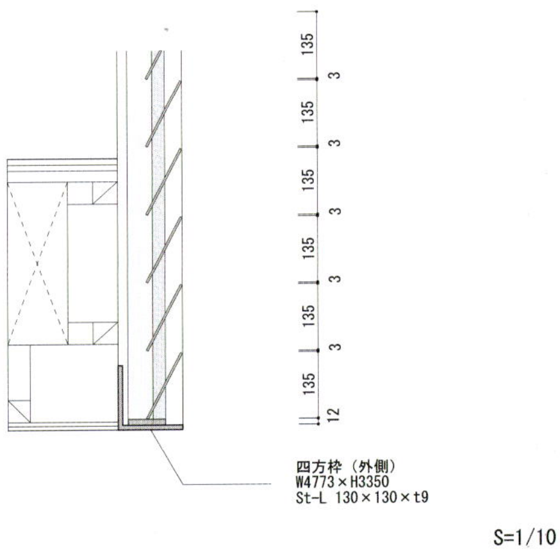
四方枠（外側）
W4773×H3350
St-L 130×130×t9
S=1/10

| 117sq.m. | Brick & Steel |

Gate House
by Henning Stummel Architects

Location _ 1 Smugglers Yard, Devonport Road, W12 8HU London
Lead Architects _ Henning Stummel & Eliana Sousa
Engineering _ Michael Hadi Associates
Gross Built Area _ 117m^2
Photography _ Henning Stummel & Rachael Smith

SMUGGLERS
YARD
W12

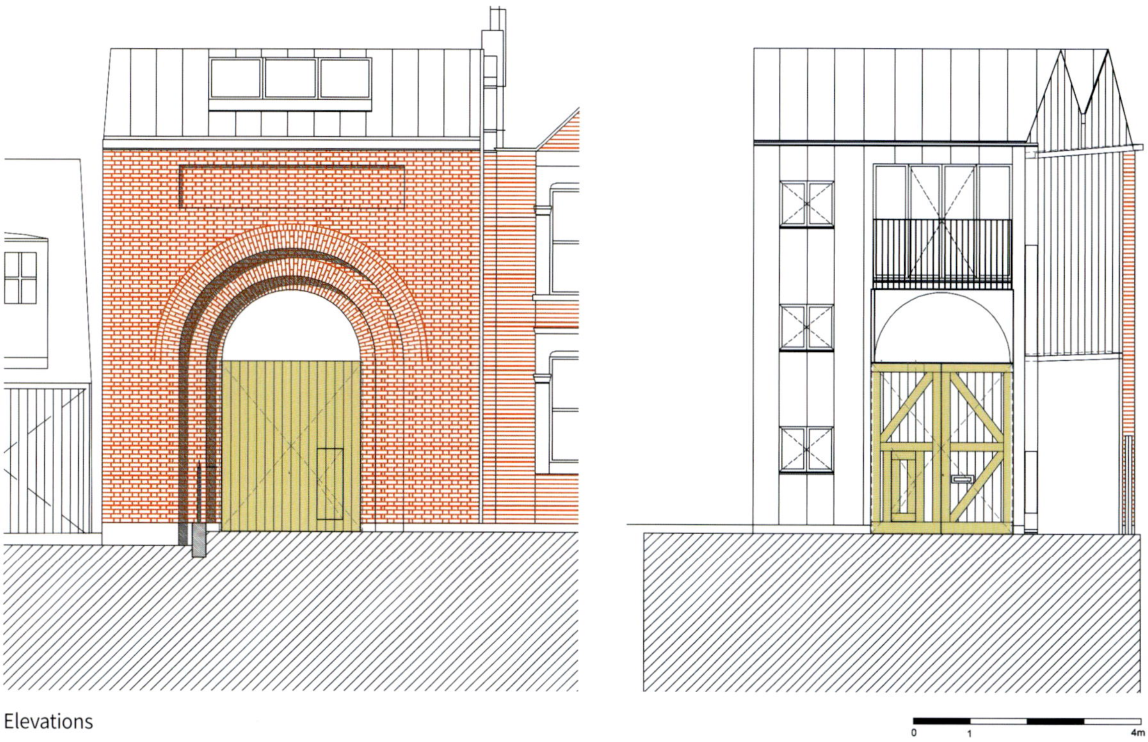

Elevations

The building above an access road creates a new site. The retained passage leads off a typical Victorian street and accesses the Tin House. The surrounding street facades are brick and a load bearing brick arch within a building that respects the building lines and parapet heights fills an unattractive gap and provides an additional new dwelling. The brick structure you see from the street is in fact only two meters deep. The remaining part of the building is a highly insulated timber structure and walls and roof are clad in a standing seam colour coated metal to match the 'Tin House'. From the outside, there are no clues as to what you will find inside. The element of surprise and discovery is important in how you experience this building. The entrance door is within the beautifully executed brick archway and leads into a double height space. Two runs of stairs take you to the second floor, taking you past a modest bedroom at ground level and a bathroom at first. These floors have low ceiling heights, but the light from the shed roofs of the generous top lit reception room guides you upstairs. This reception room, with no windows onto the street, comes as a complete surprise but has the benefit of a lovely west facing balcony, a compact kitchenette and a cozy cast iron wood burner. The valley roof reduces bulk and offers a discreet location for PV panels. This is a highly efficient house that uses air source heating and received an EPC A-rating.

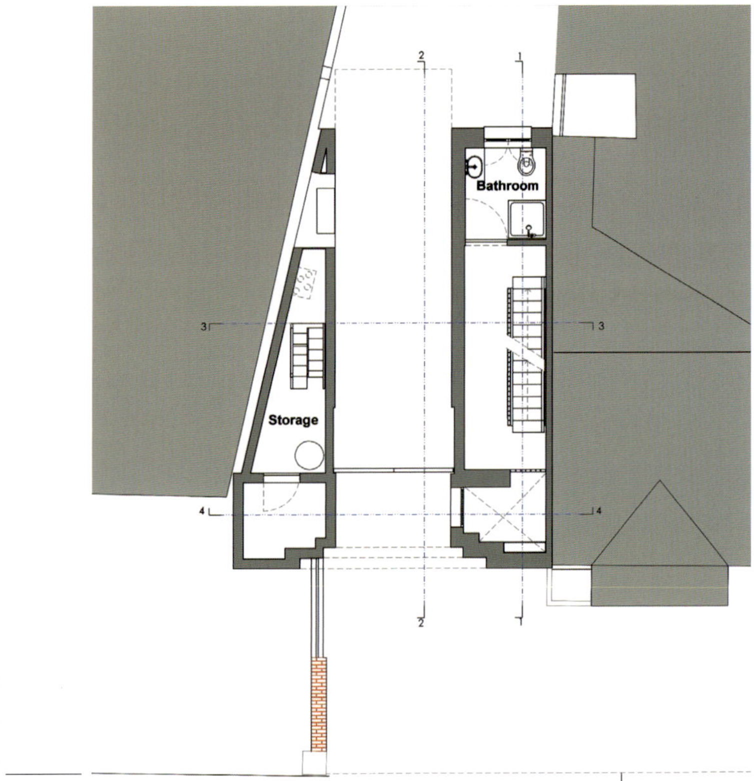

 First Floor

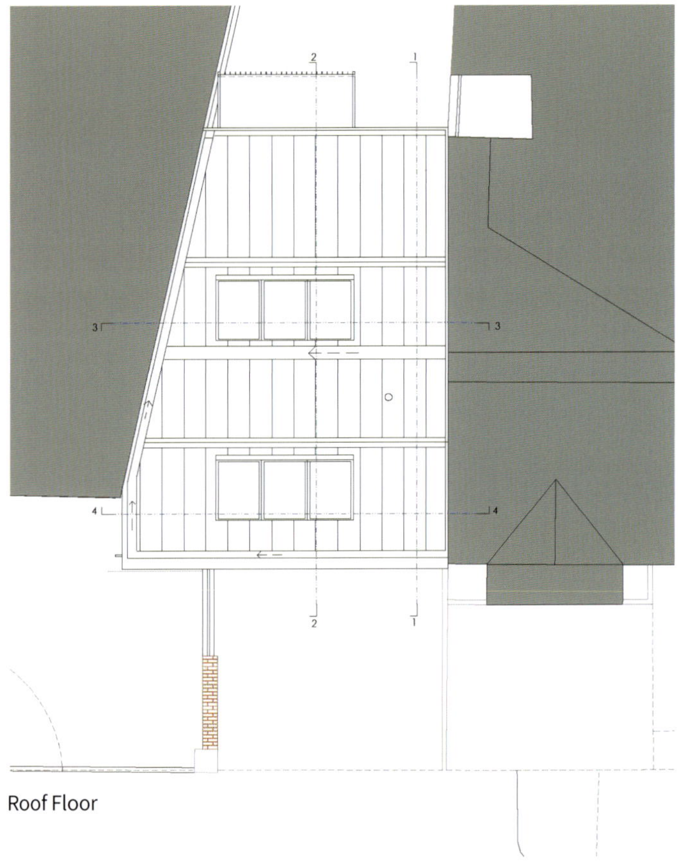

Roof Floor

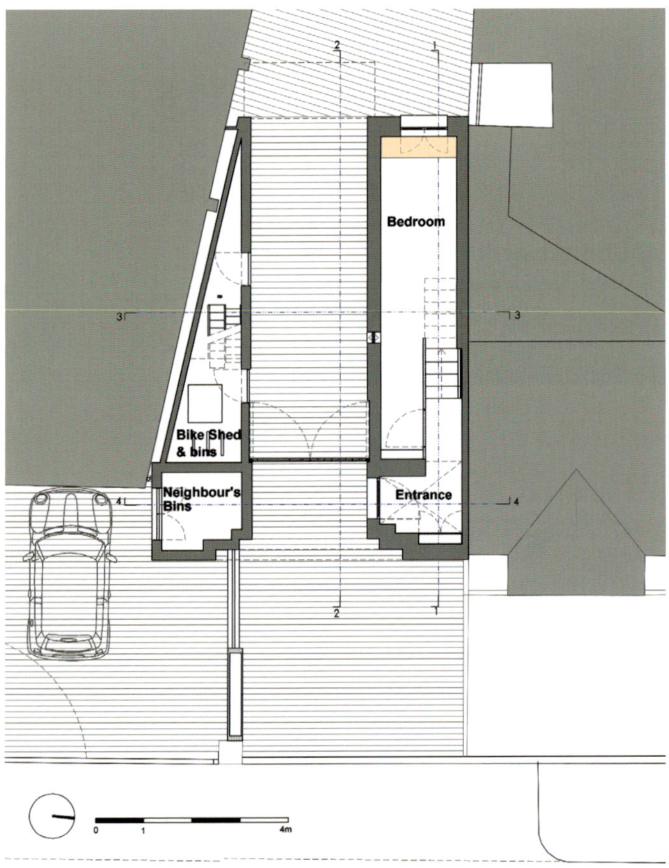

Ground Floor

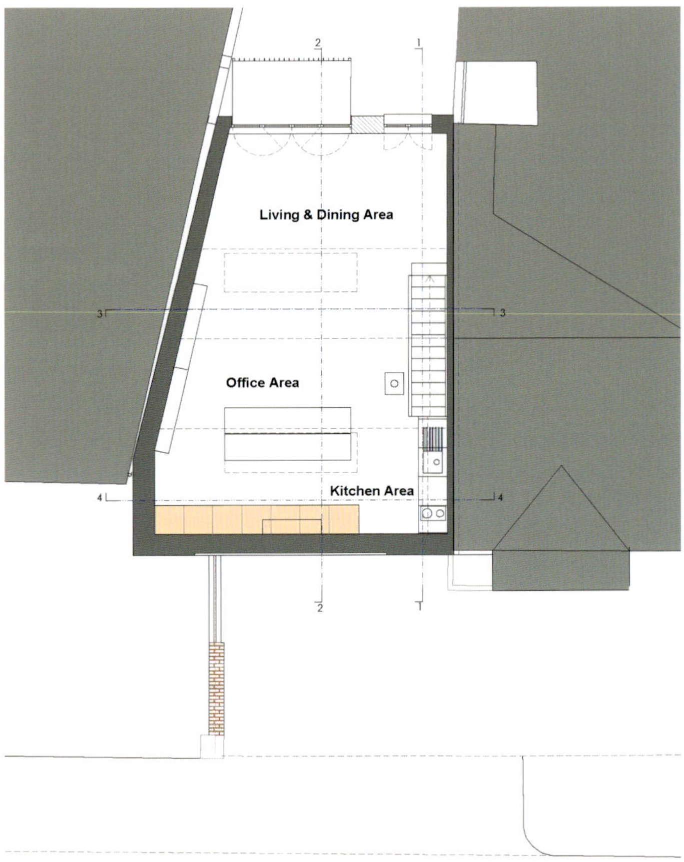

Second Floor

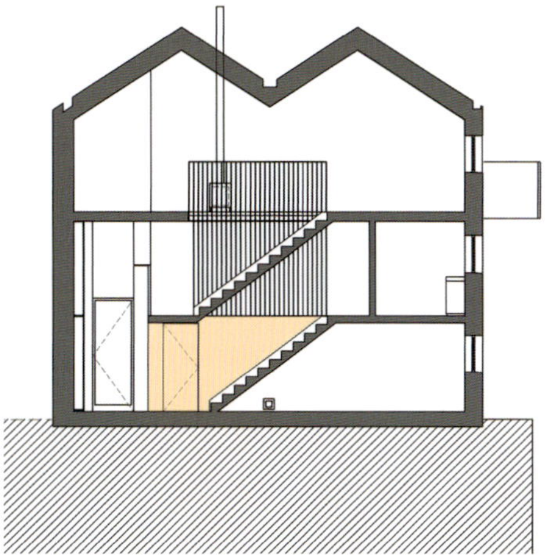
Section 1

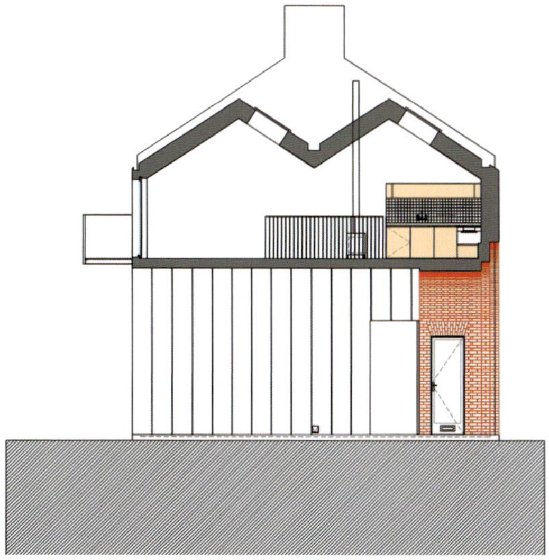
Section 2

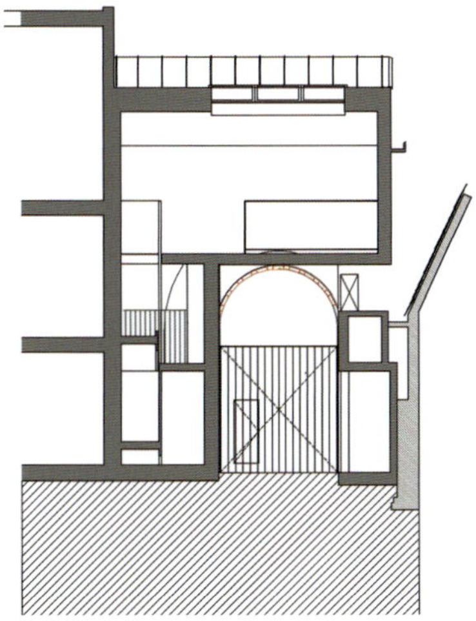

Section 3

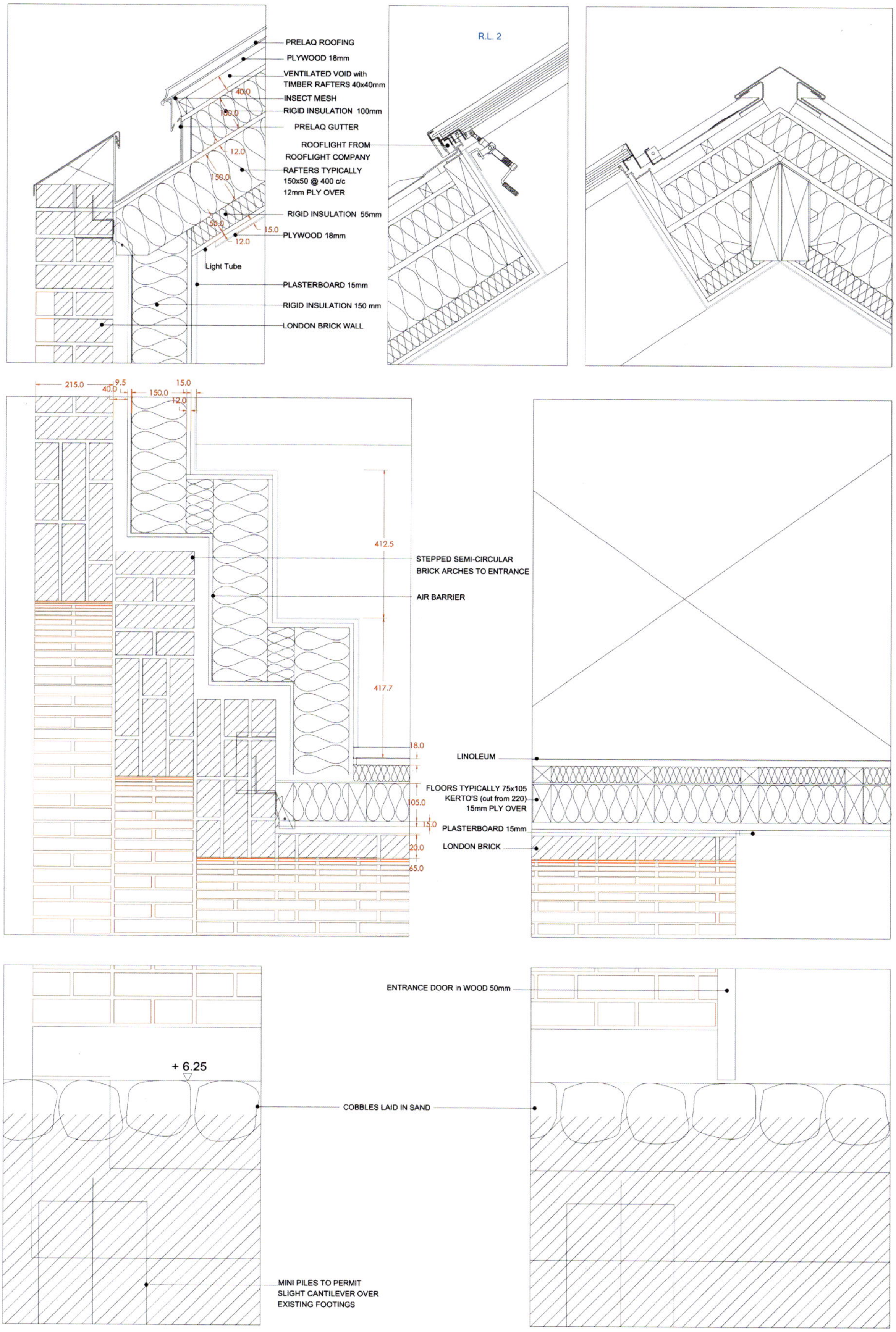

Gate House

Detail of the entrance arch and roof

0 0.1 0.2 0.5m

1:10 @ A3

ARCHITEKTURPOSITIONEN
HENNING STUMMEL
MIND THE GAP

| **118.46**sq.m. | **Wood** |

House in Shimomuraki

by Aki Hamada Architects

Architects _ Aki Hamada Architects (Aki Hamada, Takeshi Tanabe)
Location _ Toyama, Japan
Site Area _ 195.15m^2
Building Area _ 62.93m^2
Total Floor Area _ 118.46m^2
Structure _ Konishi Structural Engineers (Yasutaka Konishi, Noboru Enshu)
Photography _ Takumi Ota

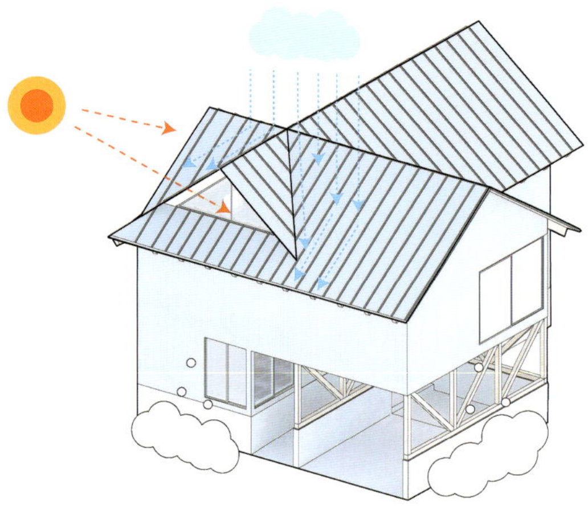

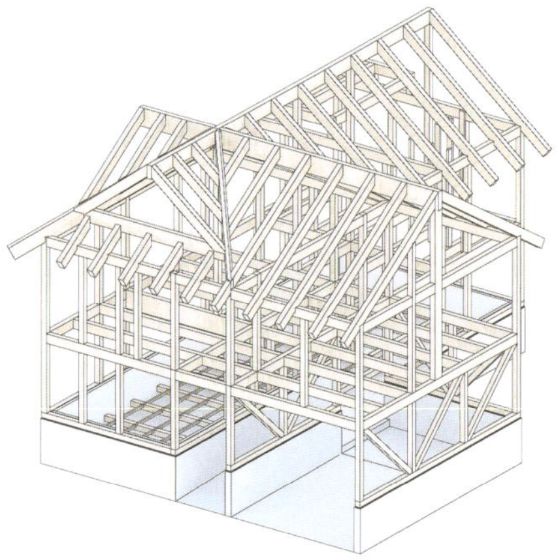

The site is in a community in Toyama with a houses and shops sprawling in a relatively relaxed manner. There is a playground in the neighboring area on the southeastern side, and the Tateyama mountain range can be seen beyond that. We placed an L-shape volume like the chumon-zukuri (a traditional L-shaped style house) inside the trapezoidal-shaped site, and arranged a child room and garage that can be used as a future shop facing a front road on the west side. Considering the snowfall, taking the distance to the north side neighbor, as well as setting the center of living space on the second floor on the southeastern park side, enable the house to be open to the park side while securing privacy.

Because this area has heavy rainfall and snowfall, we made two orthogonal gable ridge penetrate to divide the roof so that the flow of rain and snow is dispersed; the gable will be the element of environmental control to incorporate light and wind. In order to create a reasonable wood structure in the snowy area, we reduced the number of components by constructing the thatched roof structure instead of the wagoya (a traditional post-and-lintel roof framing), and constructed a large space efficiently while reducing the bending stress given to the beam. The part where the two buildings intersect becomes a three-dimensional truss, under which the center of living space is created. On the first floor, by setting the rise of the foundation of the outer peripheral part to a waist height of approximately 1 meter, we reduced the stress of the column receiving the load of the upper floor and made it function as a wall to protect the house during snowfall.

As the houses before the early modern period adapted to the climate of the land and each building element was parametrically adjusted, we reconstructed the characteristic composition and custom of the house as a technical characteristic in the world today, rather than in a symbolic manner.

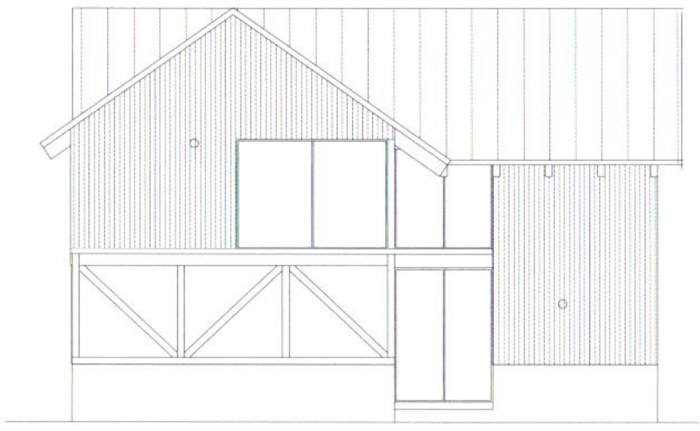

South Elevation

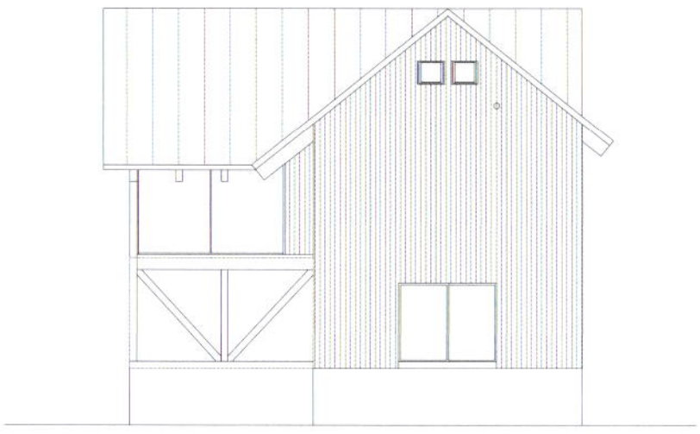

East Elevation

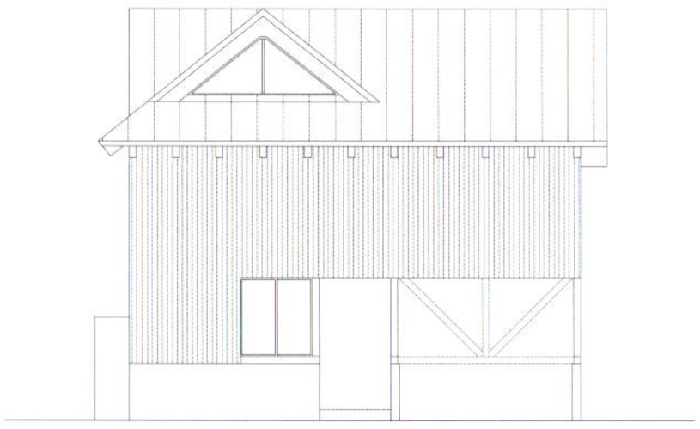

West Elevation

North Elevation

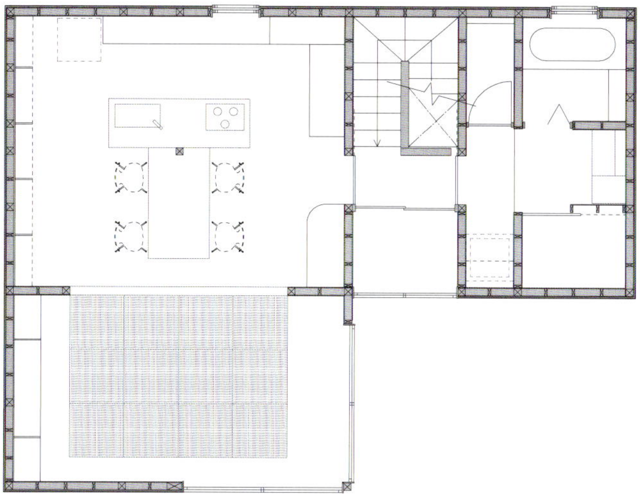

First Floor

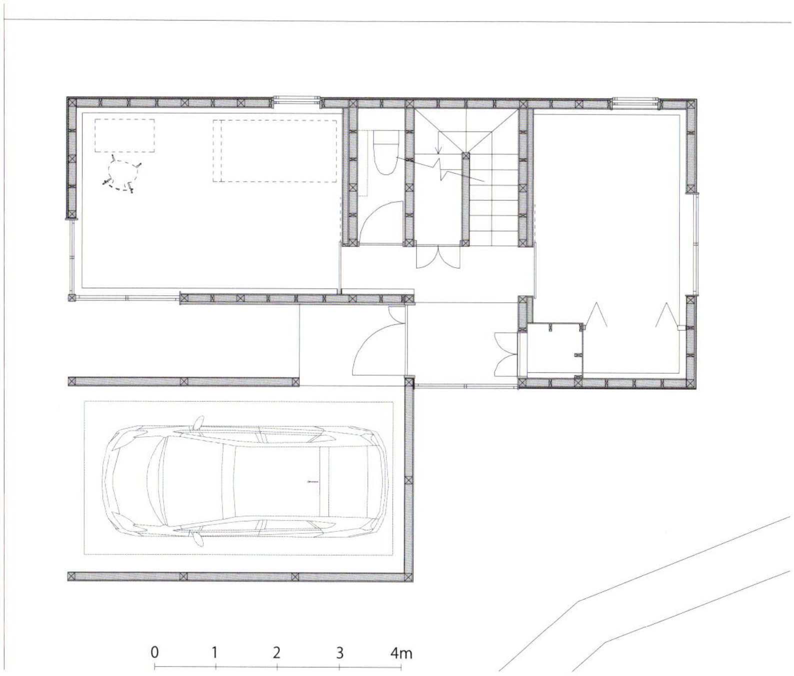

Ground Floor

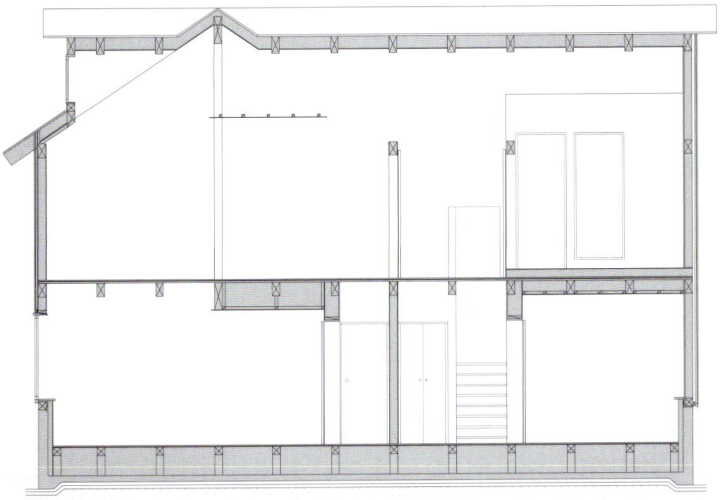

Section A

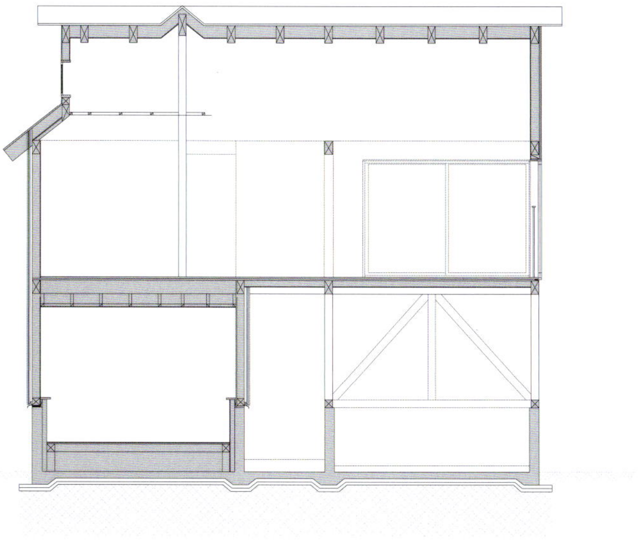

Section B

| **119.46**sq.m. | **Wood** |

House in Minoh

by Fujiwaramuro Architects

Location _ Minoh city, Osaka Pref., Japan
Lead Architects _ Shintaro Fujiwara, Yoshio Muro
Structure _ Wooden
Site Area _ 243.55m^2
Building Area _ 80.94m^2
Total Floor Area _ 119.46m^2
Photography _ Toshiyuki Yano

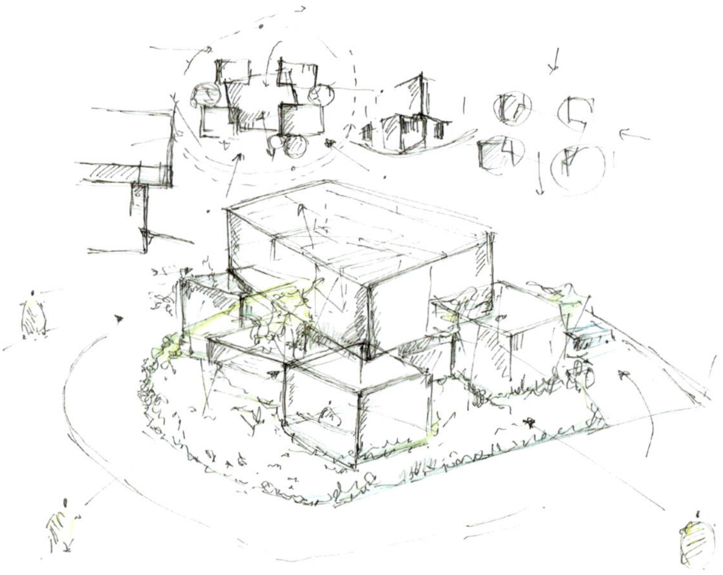

This house is located in a subdivision where rows of newly-built residences occupy large lots. The lot on which this house sits is elevated slightly above the level of the roads that surround it on three sides, and passersby are able to see all sides of the house.

The design responds to these conditions with a structure composed of stacked boxes whose appearance changes depending on the location from which they are viewed. Light enters the building from many different angles, including from clerestory windows encircling all four sides, from slit windows, and from large windows on the south side that are created by the building's exterior shape.

The large windows face onto a terrace that is surrounded by box-like outer walls, allowing the residents to enjoy time outside without worrying about being seen by passersby.

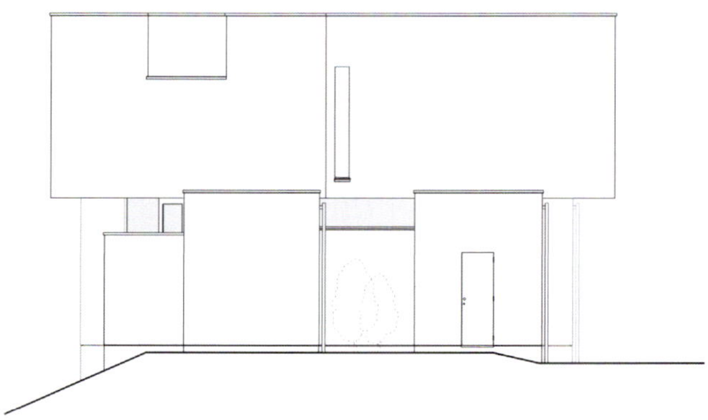

East Elevation

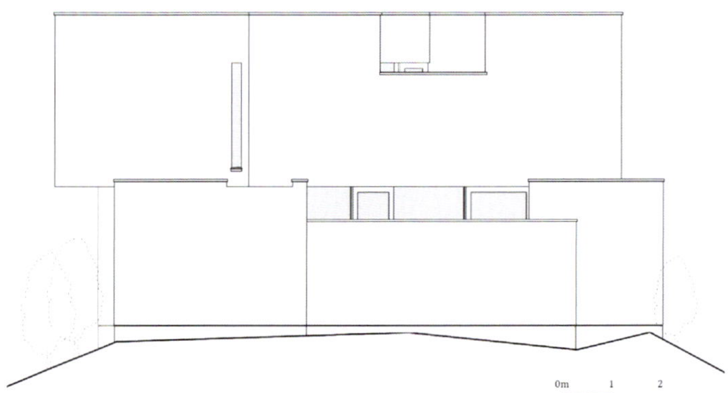

South Elevation

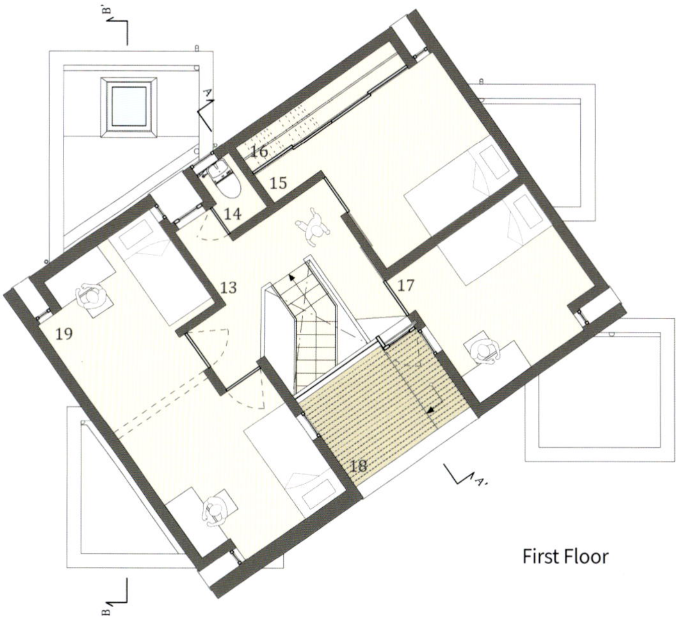

First Floor

1. Entrance
2. Shoe locker
3. Kitchen
4. Closet 1
5. Dining & Living room
6. Study room
7. Deck
8. Japanese room
9. Drying area
10. Sanitary
11. Bathroom
12. WC 1

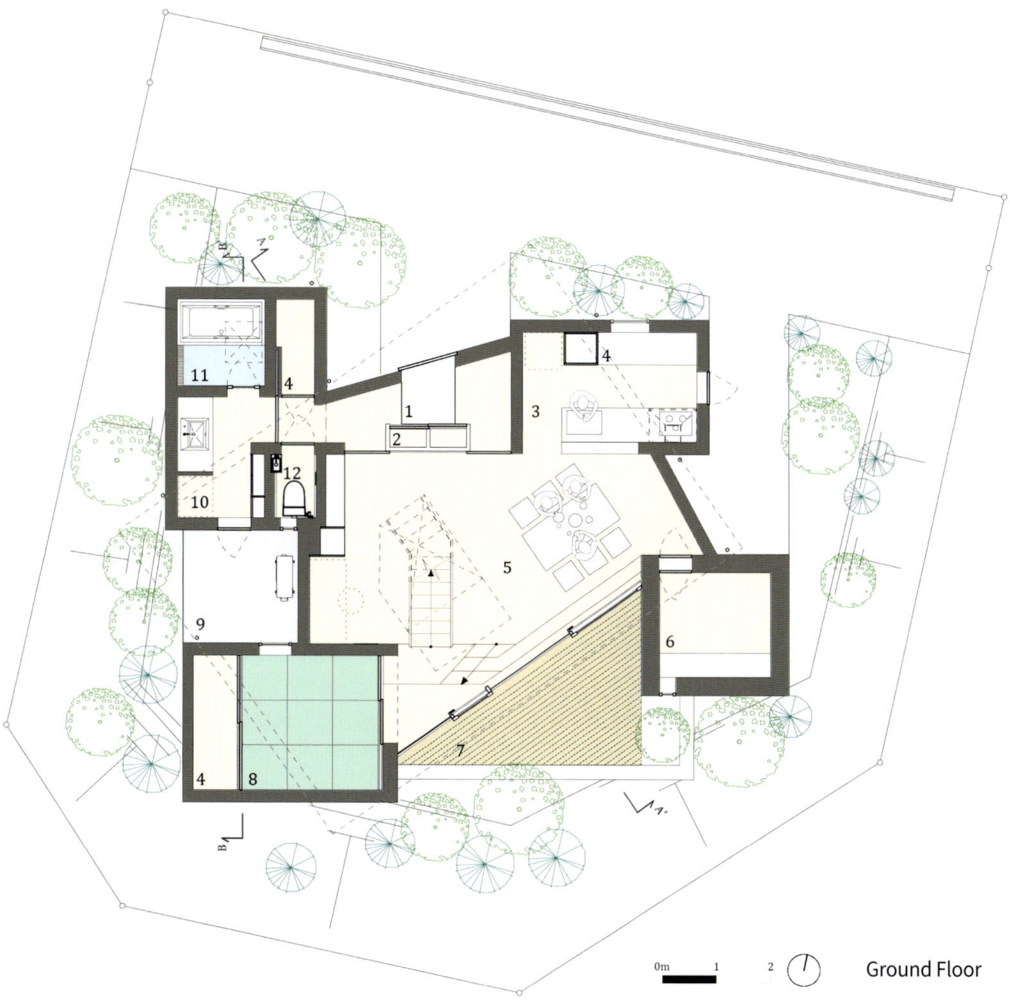

Ground Floor

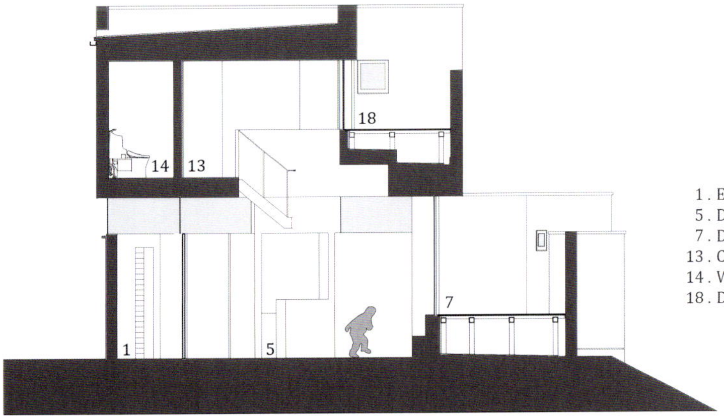

1 . Entrance
5 . Dining & Living room
7 . Dech
13 . Corridor
14 . WC 2
18 . Deck

Section A-A'

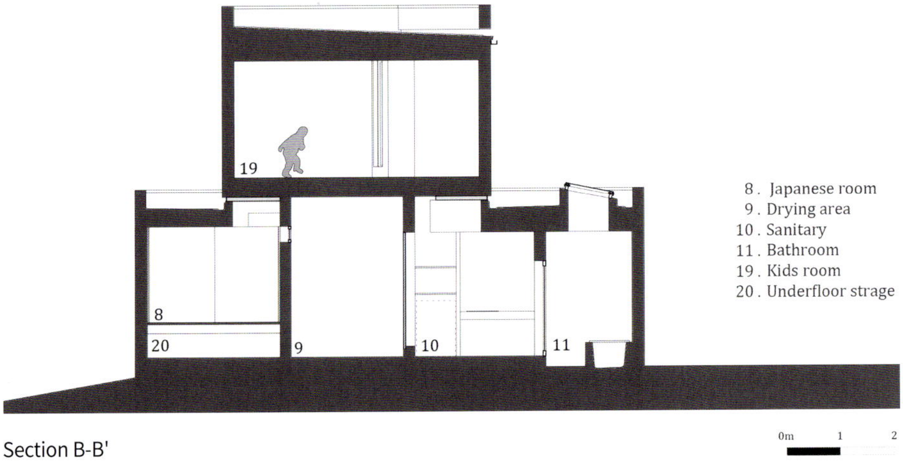

8 . Japanese room
9 . Drying area
10 . Sanitary
11 . Bathroom
19 . Kids room
20 . Underfloor strage

Section B-B'

| **122.90**sq.m. | **Wood** |

Ishibehigashi House
by ALTS Design Office

Location _ Ishibehigashi, Konan, , Shiga, Japan
Site Area _ 164.60m²
Building Area _ 76.89m²
Total Floor Area _ 122.90m²
Structure _ Wooden
Photography _ Kenta Kawamura

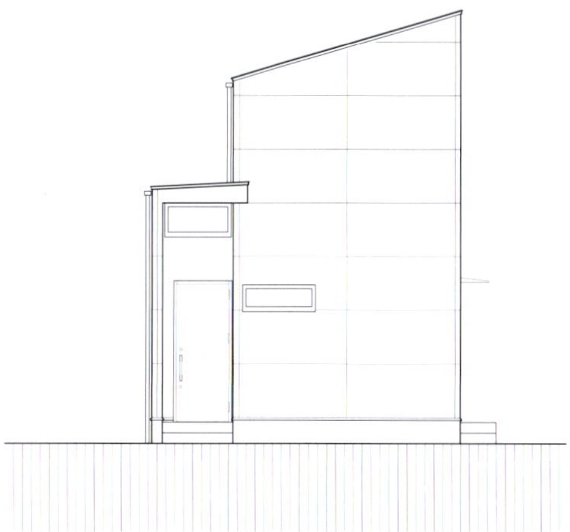

Elevation

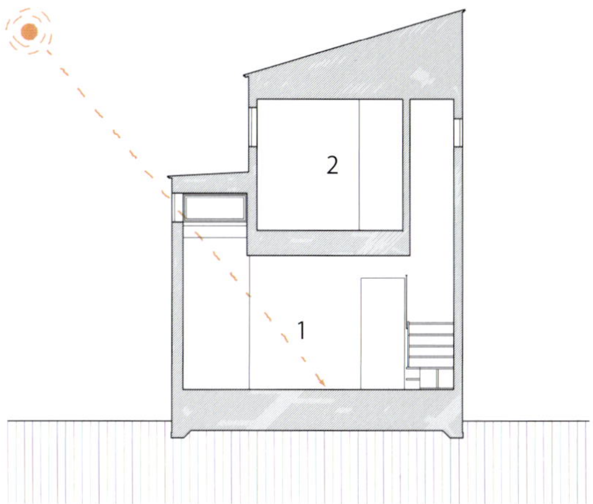

Section 1 : living 2 : bedroom

A couple decided to purchase newly developed land that was for sale. In front of their lot, there is a spreading field and a distant train is running. Although the site is surrounded by this idyllic landscape, this plot of newly developed land was elongated, having a 7-meter long width and a 23-meter depth. The houses in the area have small front areas, so it is normal to use the width of the land to its maximum extent when building a house, resulting in houses being tightly built and without much space. The area is surrounded by idyllic scenes of nature, so I felt that if the houses were located densely near each other then there would be less outside space, and that would be a shame. I tried to find a way to make more outside space, even just a little. Hence, as a way to feel the outside, we developed our concept, which was to install a garden passageway. The house is elongated, so we designed a common area with an earthen floor which was often seen in old Japanese row houses, and divided the area roughly into two parts: a passageway (the common area) and the LDK (the private area). We arranged a layout which allows the residents to access each area from the passageway. By installing an opening in the upper part of this passageway, the windows were removed windows from the field of view of the residents, so that privacy can be secured. However, as the light comes in from the top, we were able to create a space that looks like outside. As we used the same tiles as the entrance for the passageway, the interior space looked as if it is an extension of the external space. The space which looks like it's outside gone through the house. When looking up in the passageway, we can see the beams are lined up. The beams cast shadows on the LDK, and the shadows change in shape as the sun moves throughout the day. The passageway as seen from the living room is filled with soft light. The space gradually changes over time, allowing us to experience different feelings. I hope this will be a space that enriches the lives of the families living here.

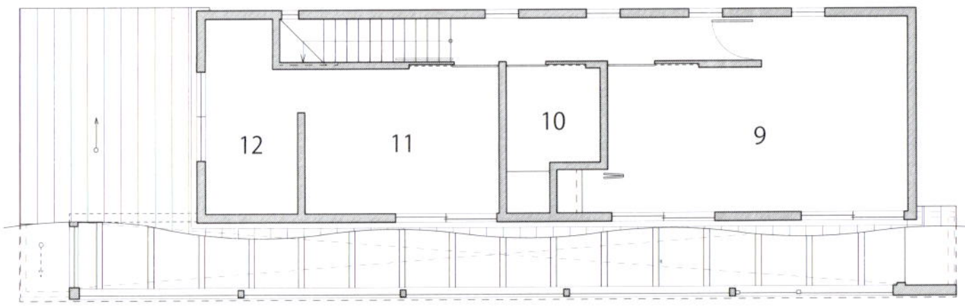

First Floor

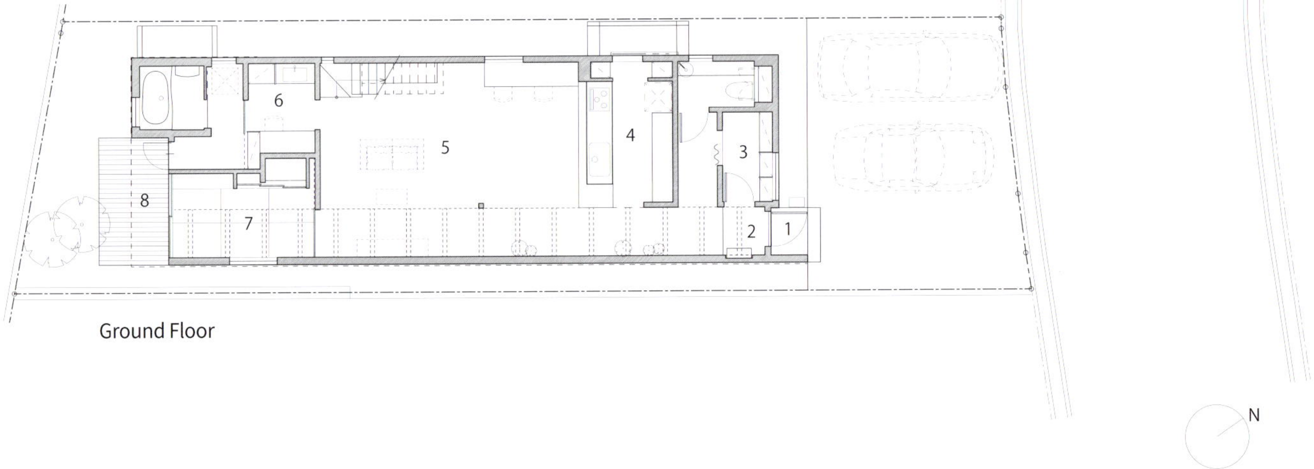

Ground Floor

1 : pouch 2 : entrance 3 : entrance storage 4 : kitchen
5 : dining 6 : washroom 7 : japanese style room 8 : wood deck
9 : children's room 10 : closet 11 : bed room 12 : WCL

| **126.44**sq.m. | **Timber** |

ACRO

by APOLLO Architects & Associates

Location _ Japan
Site Area_ 248.48m²
Building Area _82.87m²
Total Floor Area _ 126.44m²
Structure _ Timber
Scale _ Flat
Structure engineer _ Masaki structure laboratory, Kenta Masaki
Facility engineers _ Naoki Matsumoto
Lighting engineer _ Sirius lighting office, Hirohito Totsune
Constructor_ Noda House
Photography _ Masao Nishikawa

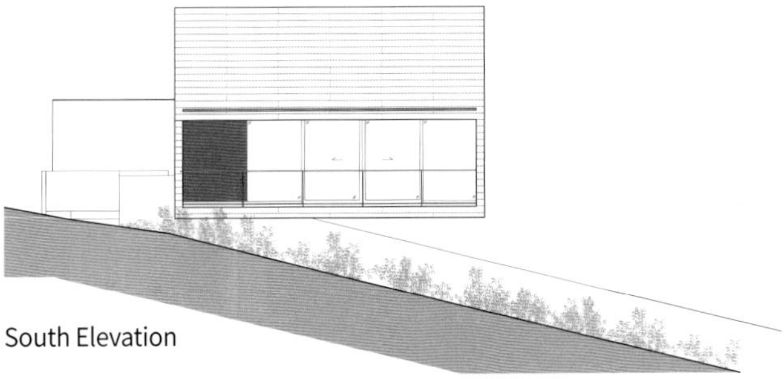
South Elevation

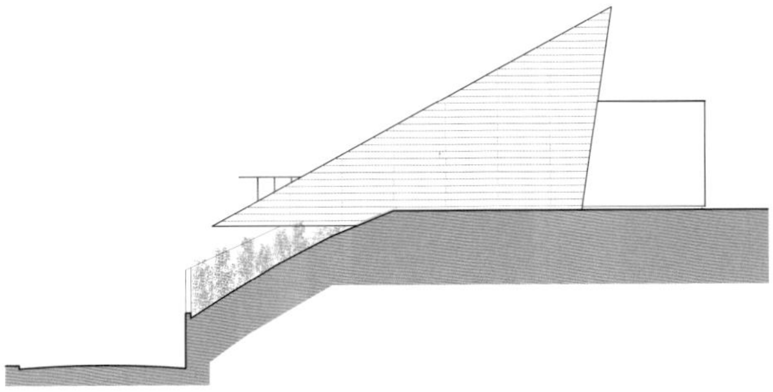
East Elevation

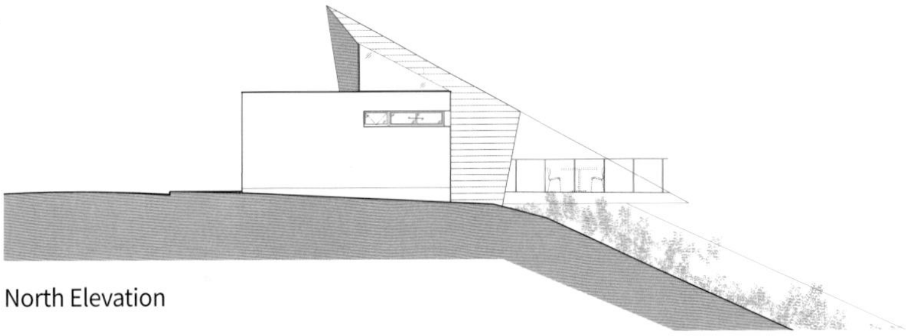
North Elevation

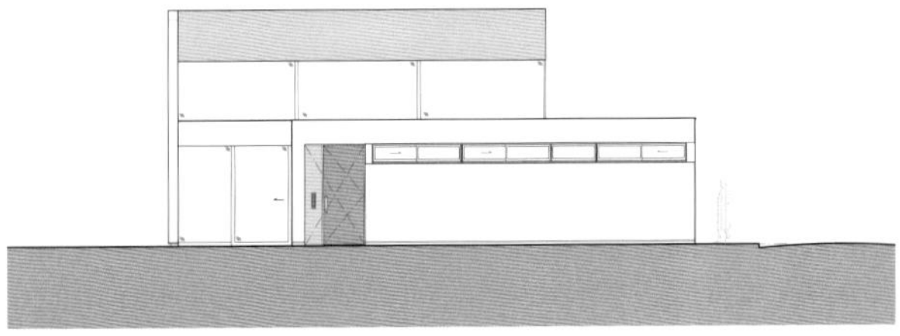
West Elevation

Located in a new residential development in Chiba prefecture, this house sits on a lot surrounded by roads on three sides. The clients requested a design that captured the expansive views to the north by taking advantage of the elevation differential between the roads and the lot.

The house is composed of a white cube and a black volume with a shed roof. This tidy yet distinctive façade gives the appearance of a two-story building, but in fact, it is a single story. Natural light pours in through the family room's clerestory windows, casting beautiful shadows on the large sloped ceiling that creates a sensation of floating. The transition from the low-ceilinged entryway to the high-ceilinged family room creates a unique rhythm that adds an element of surprise to space.

The triangular terrace connecting to the family room is paved with the same porcelain tile used inside and is at the same elevation. The doors to the terrace slide fully back, resulting in a superb intermediate space where the residents can enjoy outdoor living against the borrowed scenery of the landscape just beyond. The clients plan to use the terrace for BBQs and "glamping" with their children.

The distinct shape of the terrace jutting out over the steeply sloping lot gives the home a second façade and brings an extraordinary sense of openness to an otherwise simple residence.

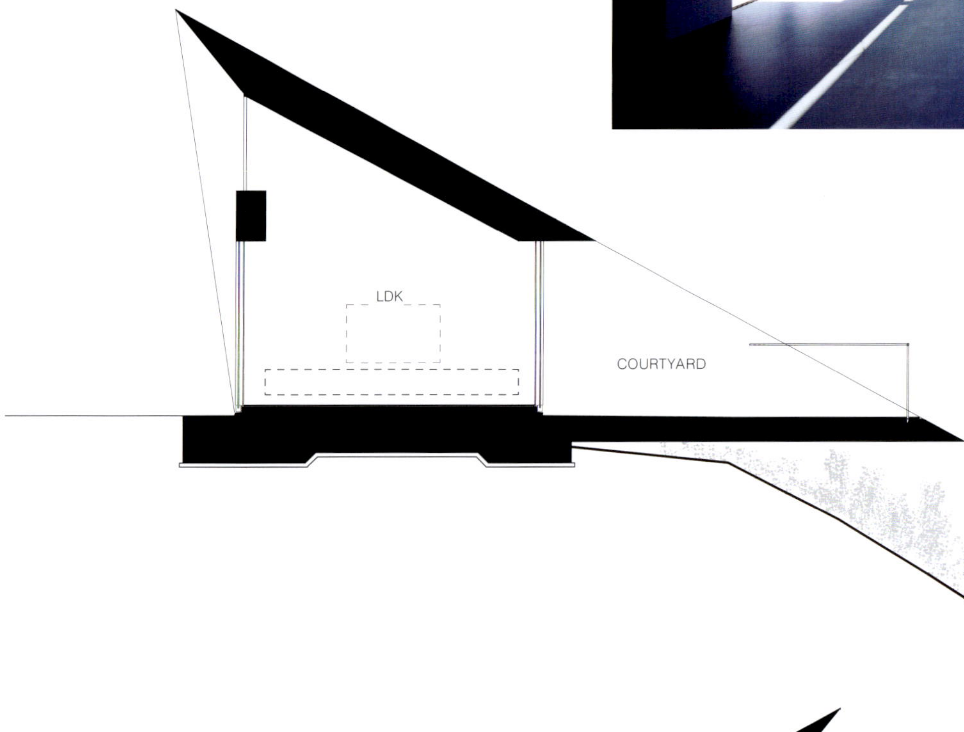

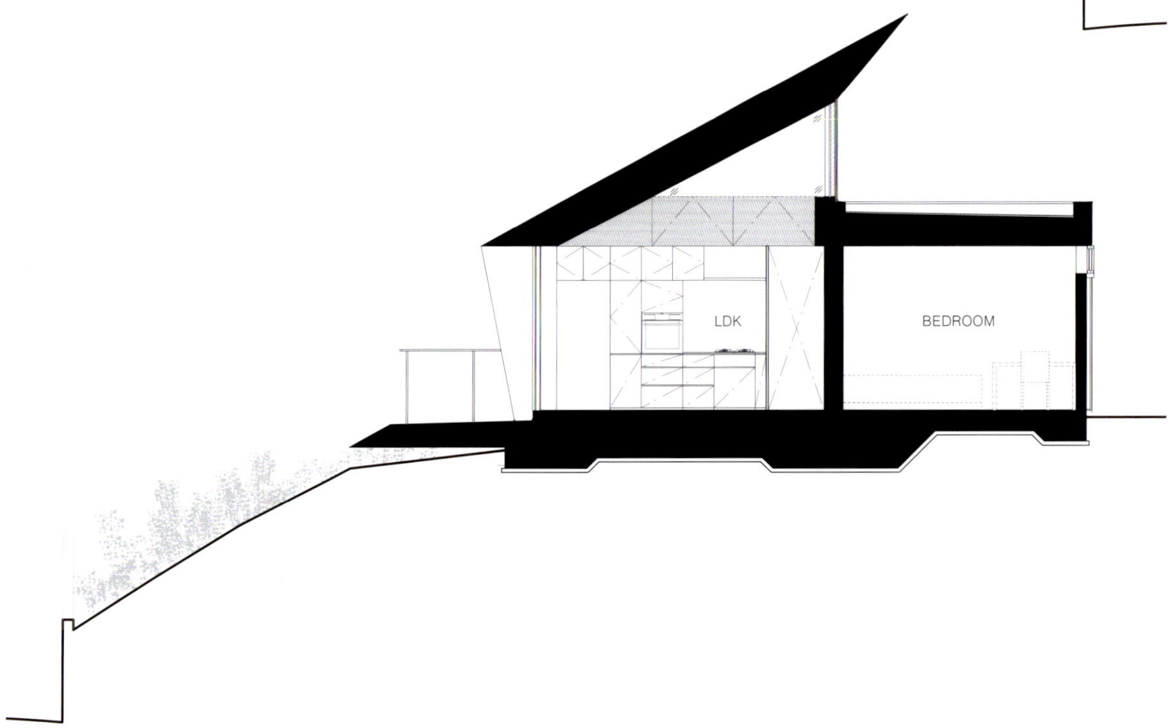

Sections

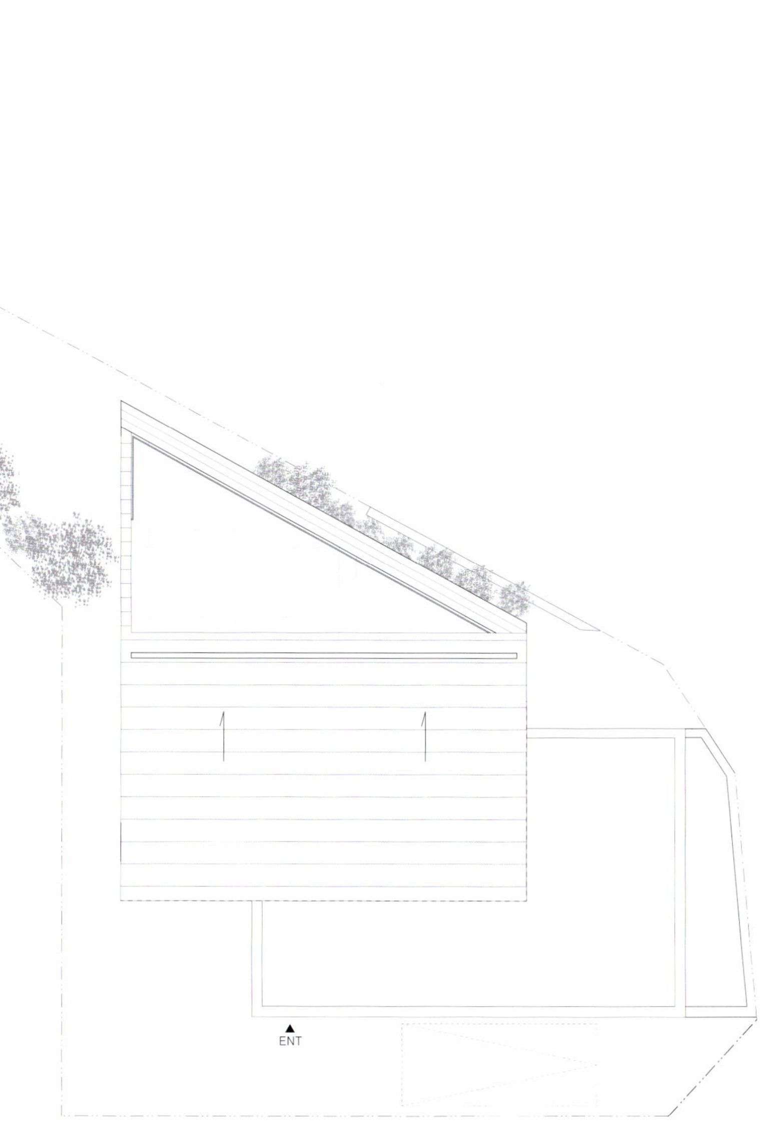

Roof Floor

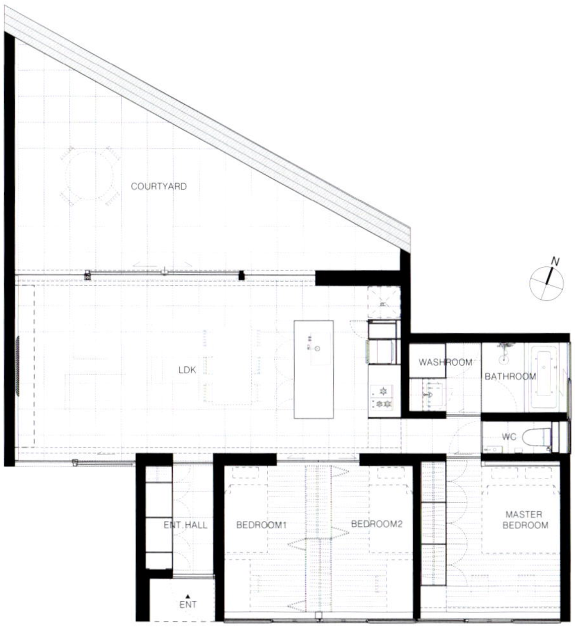

Ground Floor

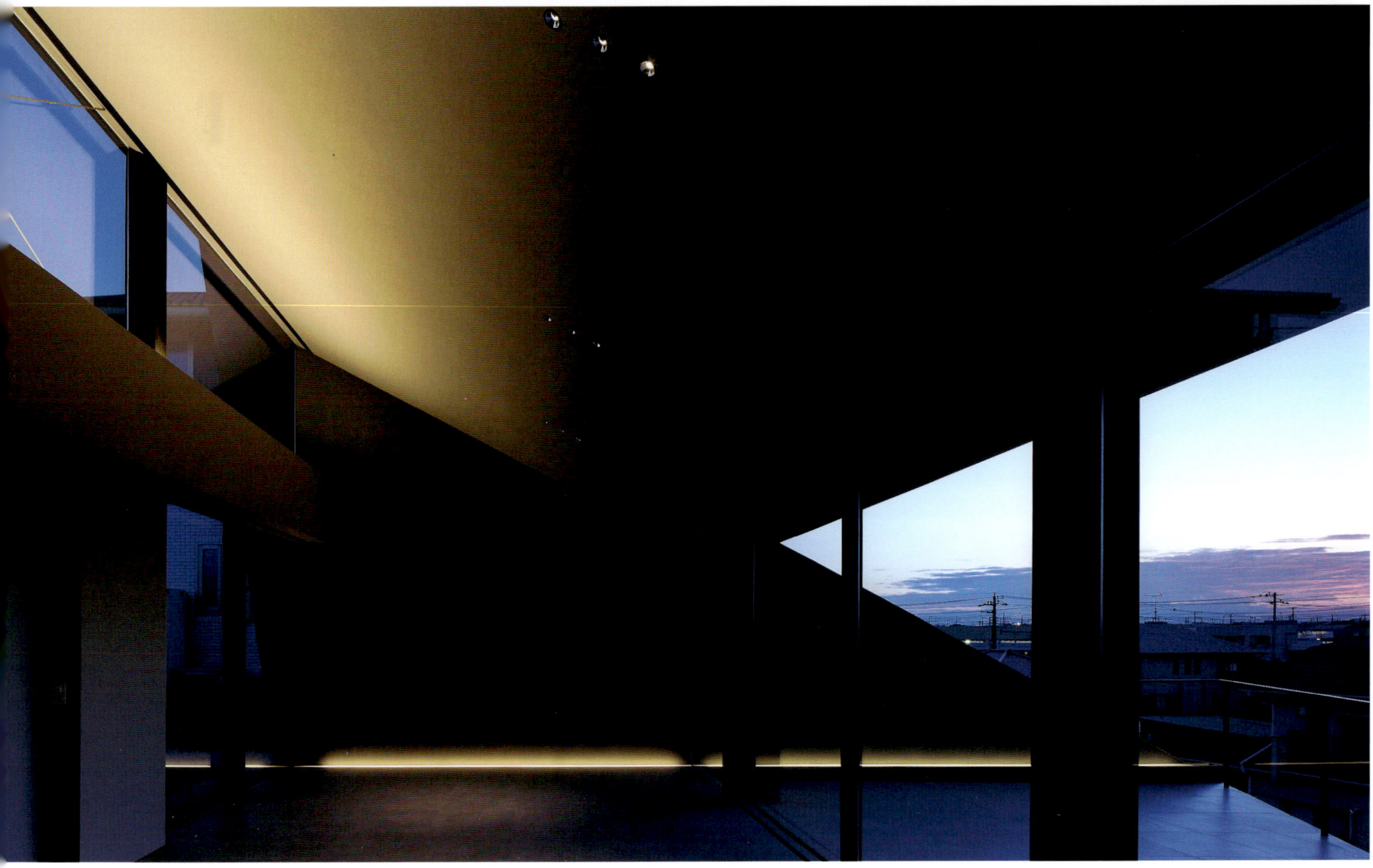

| 126.44sq.m. | Concrete |

GRID
by APOLLO Architects & Associates

Location _ Japan
Site Area_ 308.81m^2
Building Area _126.44m^2
Total Floor Area _ 126.44m^2
Structure _ RC
Scale _ Flat
Structure engineer _ Masaki structure laboratory, Kenta Masaki
Facility engineers _ Naoki Matsumoto
Lighting engineer _ Sirius lighting office, Hirohito Totsune
Constructor_ Daido Kogyo Co., Ltd
Photography _ Masao Nishikawa

The client for this project has a large art collection and wanted to start a new chapter in life with his family in a house with museum-like qualities that would take advantage of this collection. The neighborhood surrounding the site is quiet, so in order to keep noise to a minimum and maintain a sense of calm, the architect proposed a single-story reinforced-concrete structure. Like the exterior, the interior features a monochrome color scheme that serves as a good background for artwork. The building and finishing materials were also carefully selected, giving the spaces an austere atmosphere.

The layout is extremely simple, with a family room in the center with an adjacent courtyard, and a master bedroom, a children's room, and a hobby room occupying the wings on either side of the courtyard. While this symmetry gives the house a Western-style formality, the design also disrupts the symmetry in classic Japanese style by including steps and a private courtyard on only one side. The family room facing the courtyard has a high ceiling and clerestory windows on the north and south sides. The natural light that pours through these windows is partially blocked by eaves that extend out 5.9 feet (1.8 meters) and have thin outer edges that give the building a light, floating feeling. The room's coffered ceiling, divided into a 3.9- foot (1.2-meter) square lattice, resembles the architecture in temples and sukiya-style structures, which are inspired by the elegance of the classic teahouse. In addition to serving the structural purpose of supporting the weight of the roof, the ceiling turns the building itself into a piece of minimalist artwork.

By integrating the client's meticulously curated collection of artwork and the designer's furniture and lighting into the architecture, the spaces express individuality without being overly busy. As a result, the building becomes a house with museum-like qualities, rather than a museum in the shape of a house.

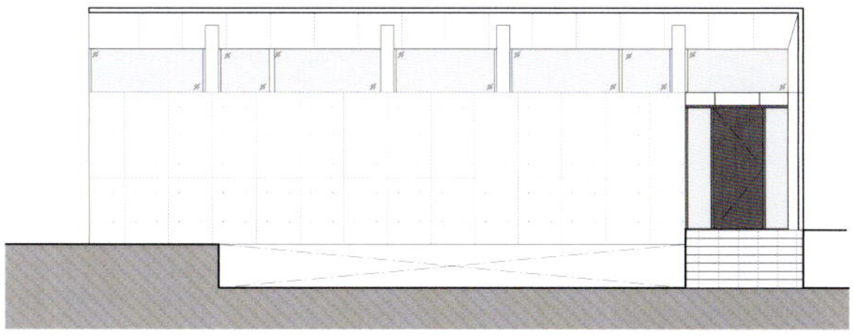

North Elevation

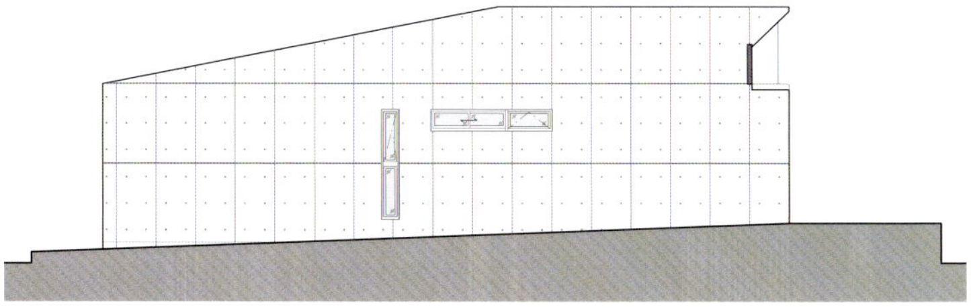

East Elevation

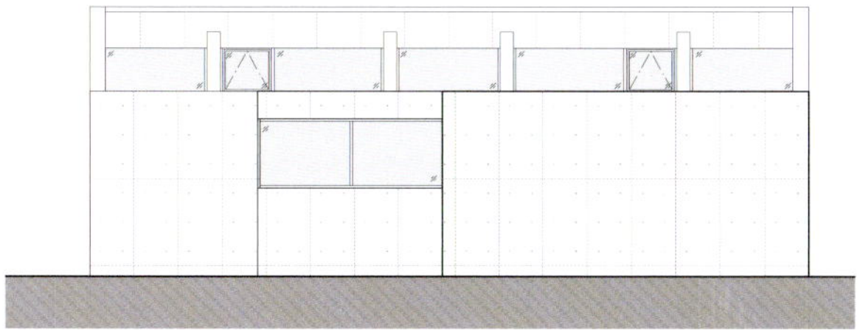

South Elevation

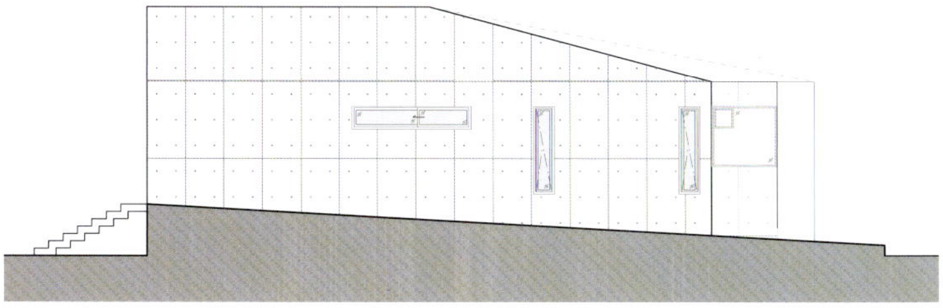

West Elevation

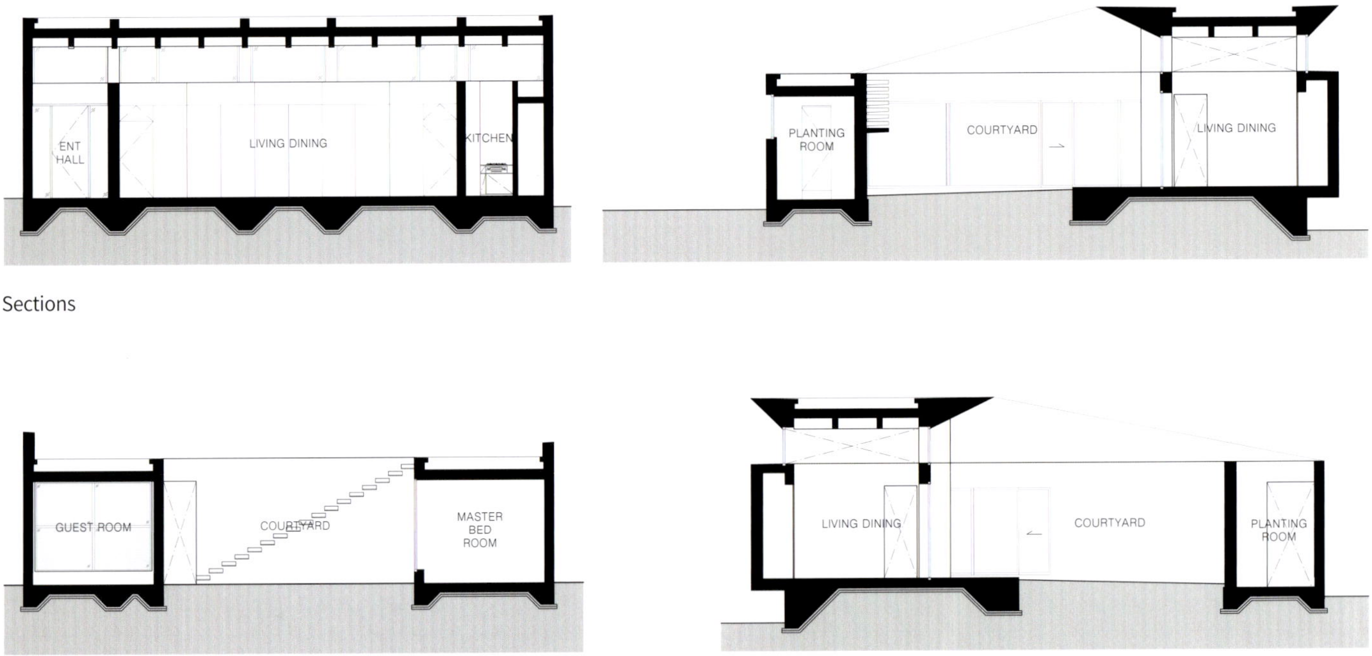
ENT HALL
LIVING DINING
KITCHEN
PLANTING ROOM
COURTYARD
LIVING DINING
GUEST ROOM
COURTYARD
MASTER BED ROOM
LIVING DINING
COURTYARD
PLANTING ROOM

Sections

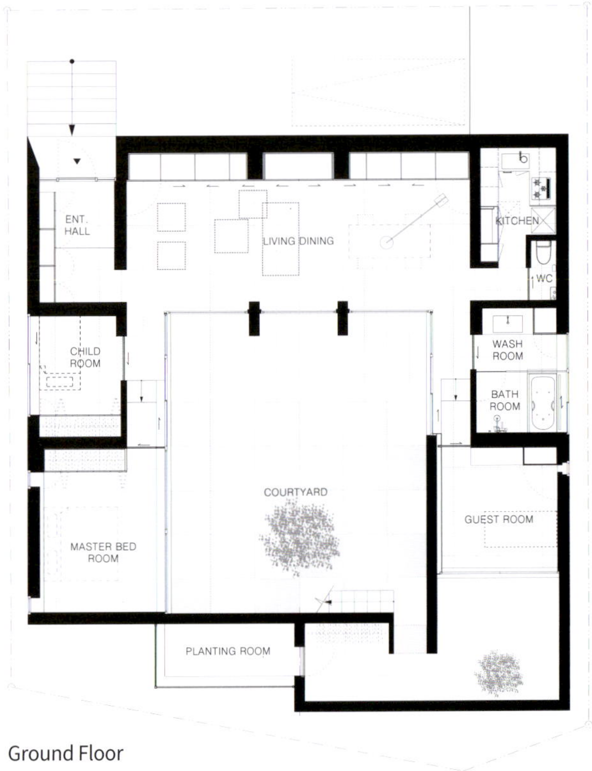

Ground Floor

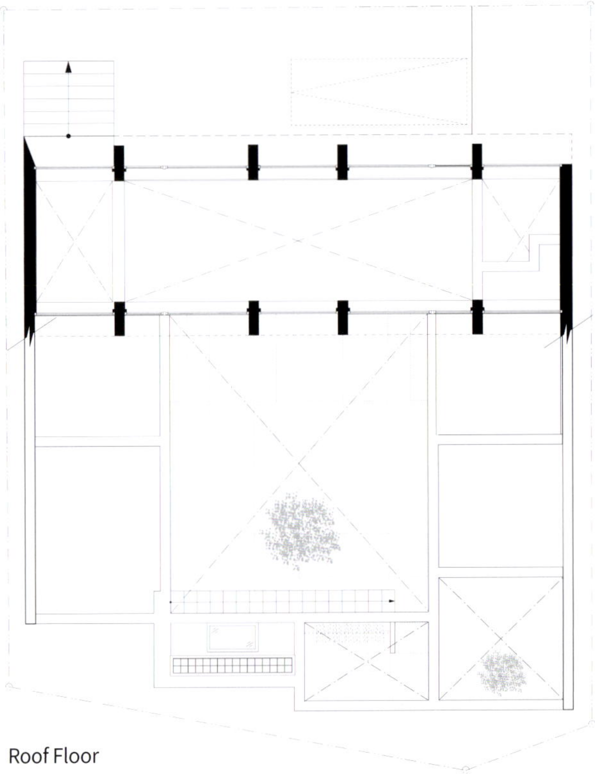

Roof Floor

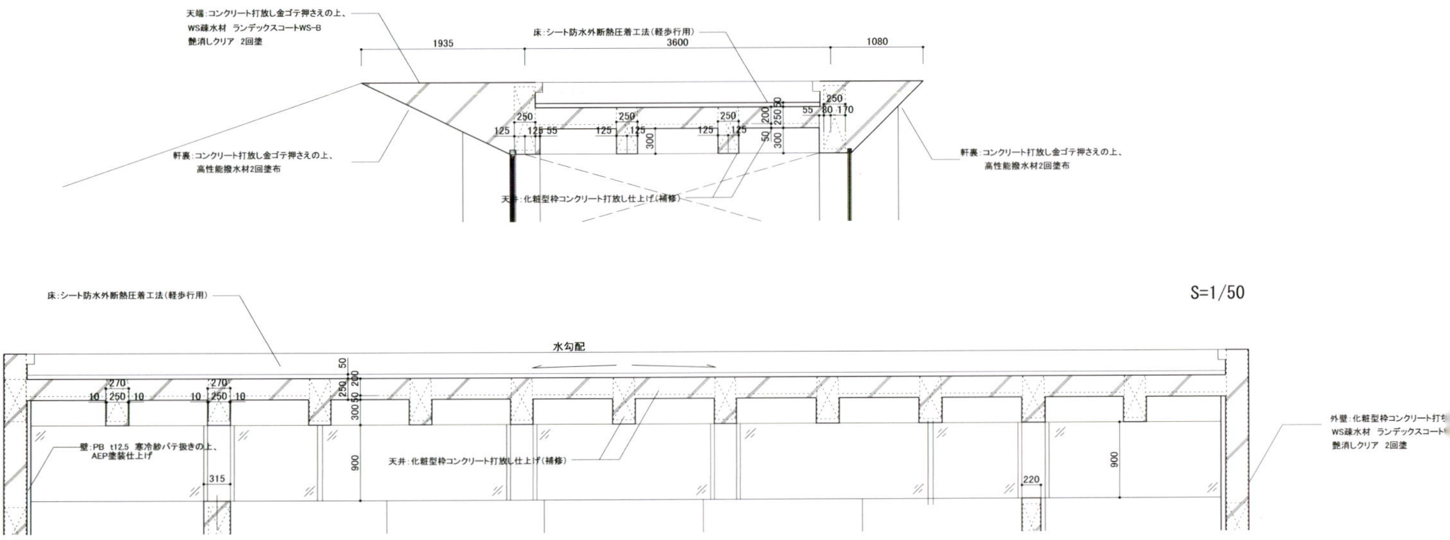
天端：コンクリート打放し金ゴテ押さえの上、
WS繊水材 ランデックスコートWS-B
艶消しクリア 2回塗
床：シート防水外断熱圧着工法（軽歩行用）
1935
3600
1080
軒裏：コンクリート打放し金ゴテ押さえの上、
高性能撥水材2回塗布
天井：化粧型枠コンクリート打放し仕上げ（補修）
床：シート防水外断熱圧着工法（軽歩行用）
S=1/50
水勾配
壁：PB t12.5 寒冷紗パテ扱きの上、
AEP塗装仕上げ
天井：化粧型枠コンクリート打放し仕上げ（補修）
外壁：化粧型枠コンクリート打
WS繊水材 ランデックスコート
艶消しクリア 2回塗

131.04sq.m. | **Timber**

Garage Hall House

by Tsukagoshi Miyashita Sekkei + Keitarchi

Location _ Tokyo, Japan
Total Floor Area_ 131.04m²
Photography _ Kenta Hasegawa

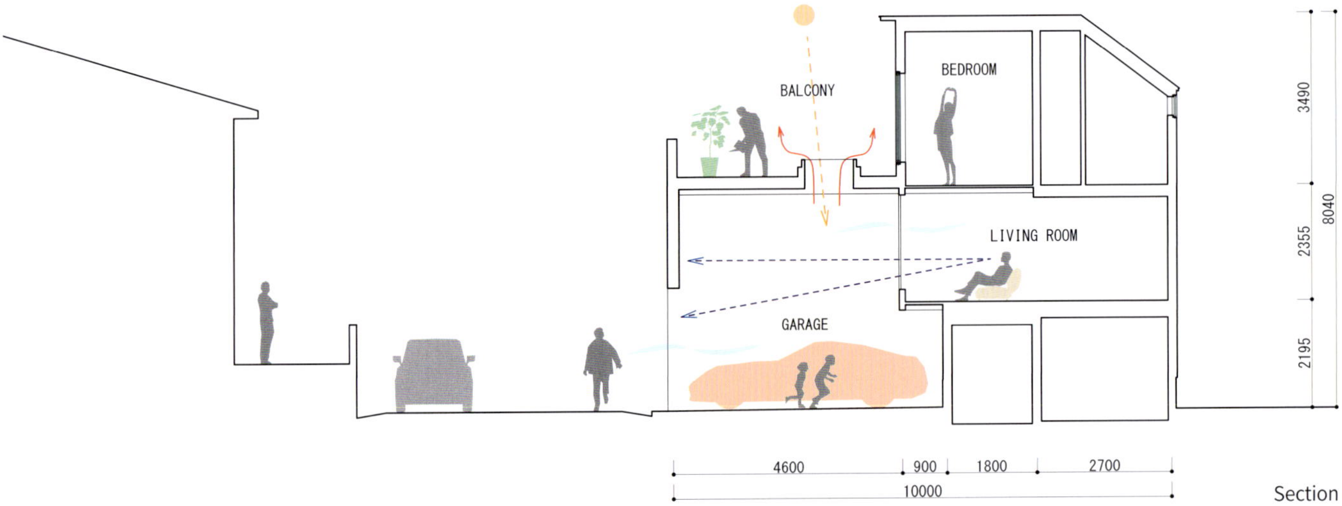

Section

This house is located in a residential area in Setagaya-city and designed for a married couple, their child and their two cars. Though this residential area is located in city center and transport there is convenient, many of the residents have cars parked in built-in car garages with shutters. Since houses in Japan are often built in narrow sites, typical houses with built-in garage have their living space back of the garage where lighting and ventilation condition is inadequate. This project aimed to reform the form of houses with a garage and to show new possibilities to them which are built repeatedly in the area. This house has a wellhole-style hall on the road side; back of the hall is a three-story living space. The hall has a wide opening to the full toward the building frontage on the roadside, openable top lights, a large high side light, and a double sliding window. This hall is used as a garage and provides an expanse of space and improve the lighting and ventilation condition of the living space. The hall works as a buffer zone between the living space and the street. The hall has a living space-like interior not typical garage-like finishing; ceiling with exposed Oregon pine beams, chandelier-like pendant illumination fixtures, dirt floor with tile-like joint. In this project, a garage which is usually considered as a negative element in a small house, is enlarged and designed as a hall to improve the living condition and expand the possibilities of how the residents live in this house. In fact, the client is decorating paintings in the hall and it has started to work as a gallery.

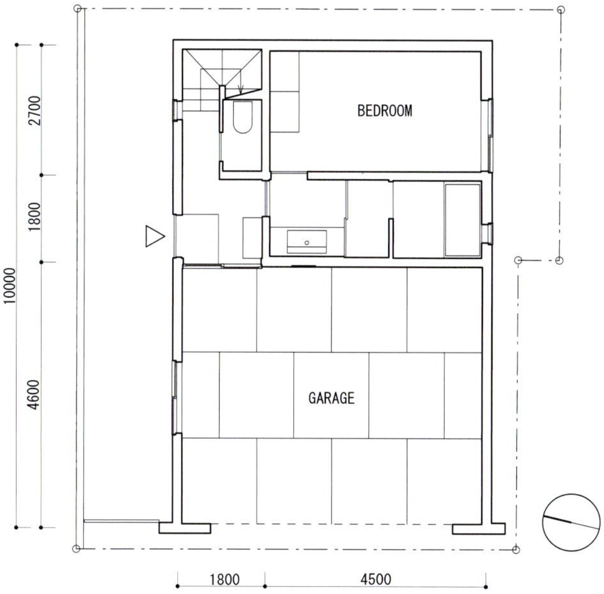

Ground Floor

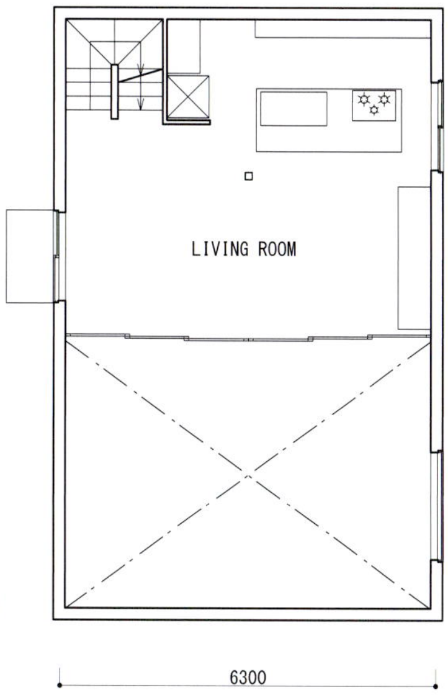

First Floor

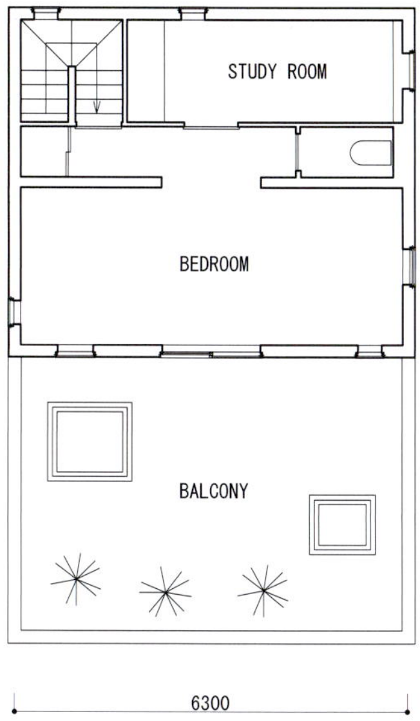

Second Floor

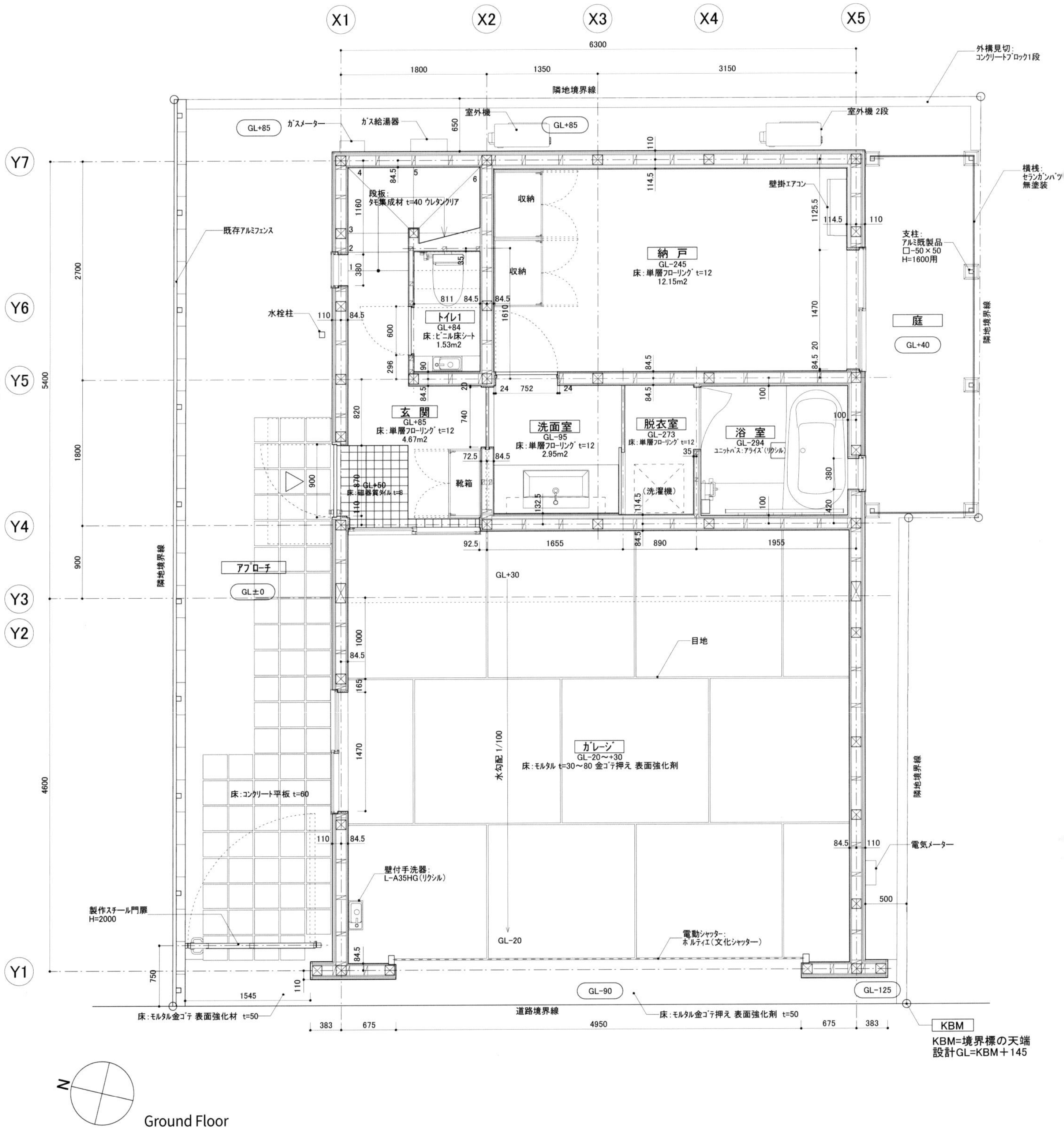

Ground Floor

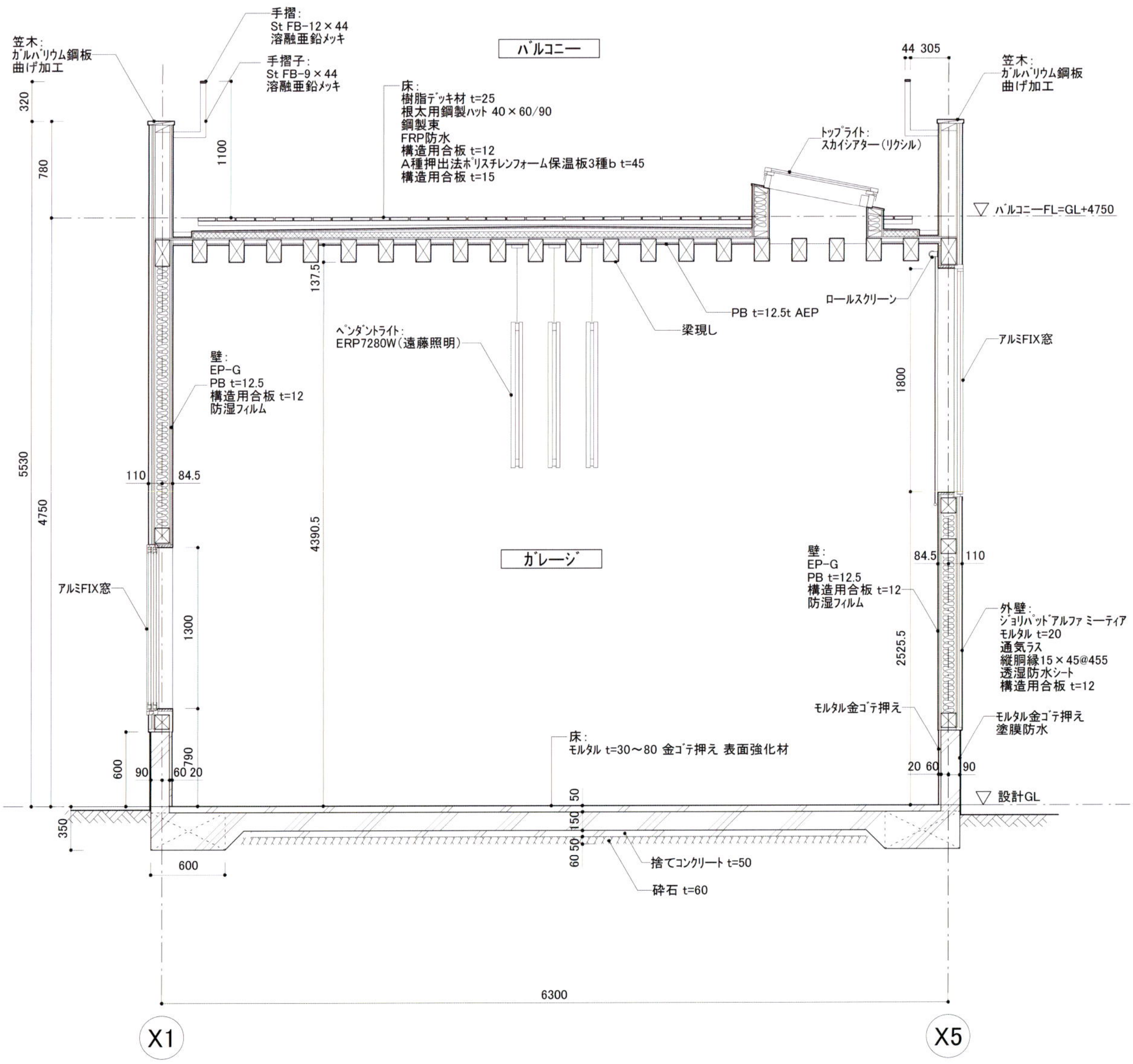

バルコニー
ガレージ
手摺：St FB-12×44 溶融亜鉛メッキ
手摺子：St FB-9×44 溶融亜鉛メッキ
笠木：ガルバリウム鋼板 曲げ加工
床：樹脂デッキ材 t=25 根太用鋼製ハット 40×60/90 鋼製束 FRP防水 構造用合板 t=12 A種押出法ポリスチレンフォーム保温板3種b t=45 構造用合板 t=15
トップライト：スカイシアター（リクシル）
バルコニーFL=GL+4750
ロールスクリーン
PB t=12.5t AEP
梁現し
ペンダントライト：ERP7280W（遠藤照明）
壁：EP-G PB t=12.5 構造用合板 t=12 防湿フィルム
アルミFIX窓
外壁：ジョリパットアルファ ミーティア モルタル t=20 通気ラス 縦胴縁15×45@455 透湿防水シート 構造用合板 t=12
モルタル金ゴテ押え
モルタル金ゴテ押え 塗膜防水
床：モルタル t=30～80 金ゴテ押え 表面強化材
設計GL
捨てコンクリート t=50
砕石 t=60
6300
X1
X5

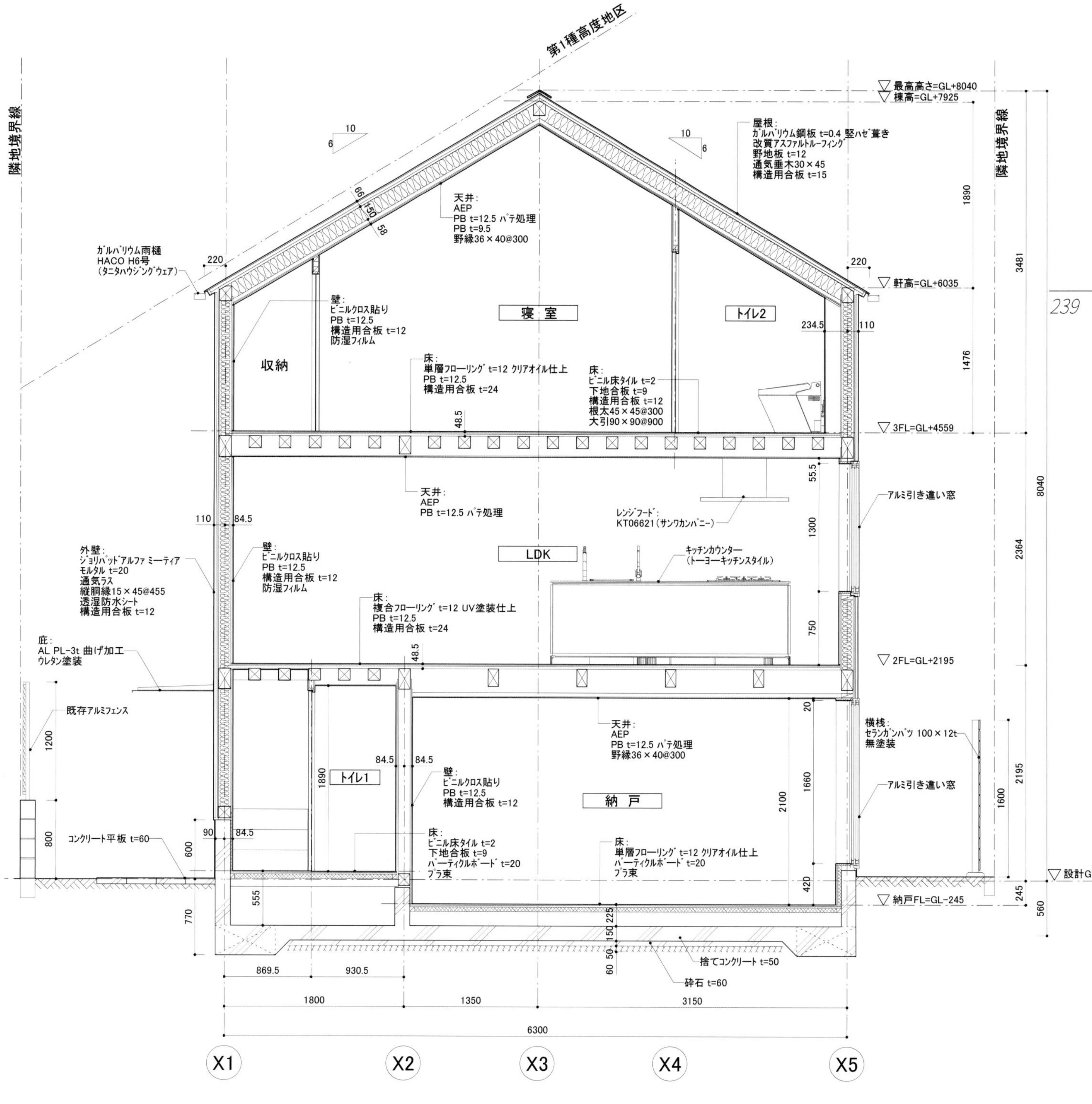

第1種高度地区
隣地境界線
最高高さ=GL+8040
棟高=GL+7925
屋根:
ガルバリウム鋼板 t=0.4 竪ハゼ葺き
改質アスファルトルーフィング
野地板 t=12
通気垂木30×45
構造用合板 t=15
天井:
AEP
PB t=12.5 パテ処理
PB t=9.5
野縁36×40@300
ガルバリウム雨樋
HACO H6号
(タニタハウジングウェア)
軒高=GL+6035
壁:
ビニルクロス貼り
PB t=12.5
構造用合板 t=12
防湿フィルム
寝室
トイレ2
収納
床:
単層フローリング t=12 クリアオイル仕上
PB t=12.5
構造用合板 t=24
床:
ビニル床タイル t=2
下地合板 t=9
構造用合板 t=12
根太45×45@300
大引90×90@900
3FL=GL+4559
天井:
AEP
PB t=12.5 パテ処理
レンジフード:
KT06621 (サンワカンパニー)
アルミ引き違い窓
外壁:
ジョリパットアルファ ミーティア
モルタル t=20
通気ラス
縦胴縁15×45@455
透湿防水シート
構造用合板 t=12
LDK
キッチンカウンター
(トーヨーキッチンスタイル)
床:
複合フローリング t=12 UV塗装仕上
PB t=12.5
構造用合板 t=24
庇:
AL PL-3t 曲げ加工
ウレタン塗装
2FL=GL+2195
既存アルミフェンス
横桟:
セランガンバツ 100×12t
無塗装
トイレ1
納戸
コンクリート平板 t=60
床:
ビニル床タイル t=2
下地合板 t=9
パーティクルボード t=20
プラ束
床:
単層フローリング t=12 クリアオイル仕上
パーティクルボード t=20
プラ束
設計GL
納戸FL=GL-245
捨てコンクリート t=50
砕石 t=60
1800
1350
3150
6300
869.5
930.5
X1
X2
X3
X4
X5
8040
3481
2364
2195

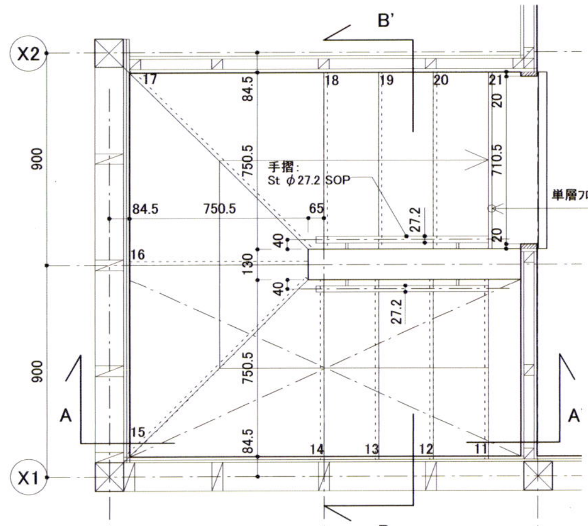

3F PLAN (S=1/

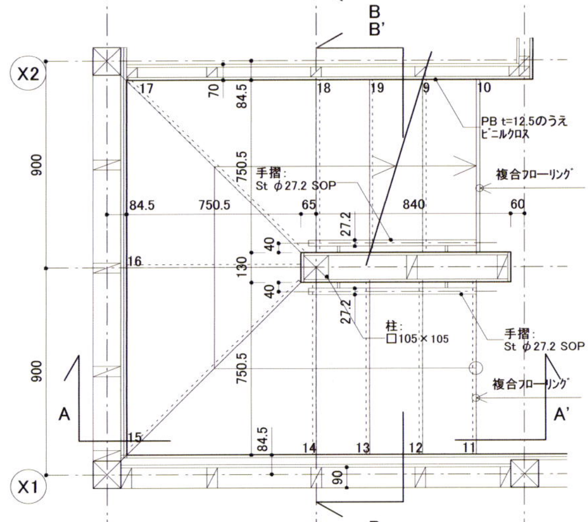

2F PLAN (S=1/

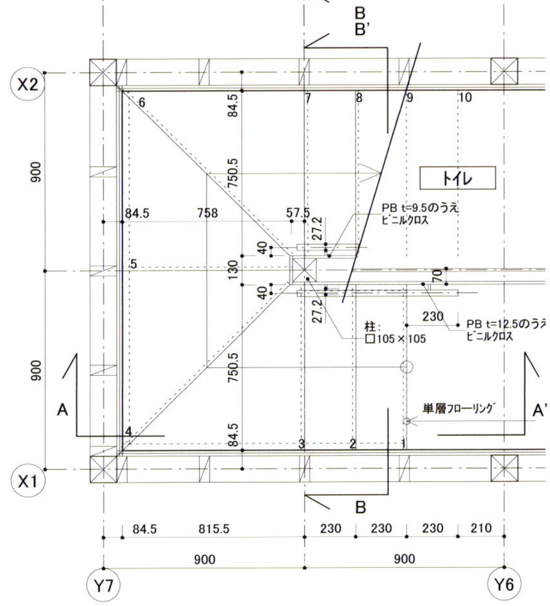

1F PLAN (S=1/

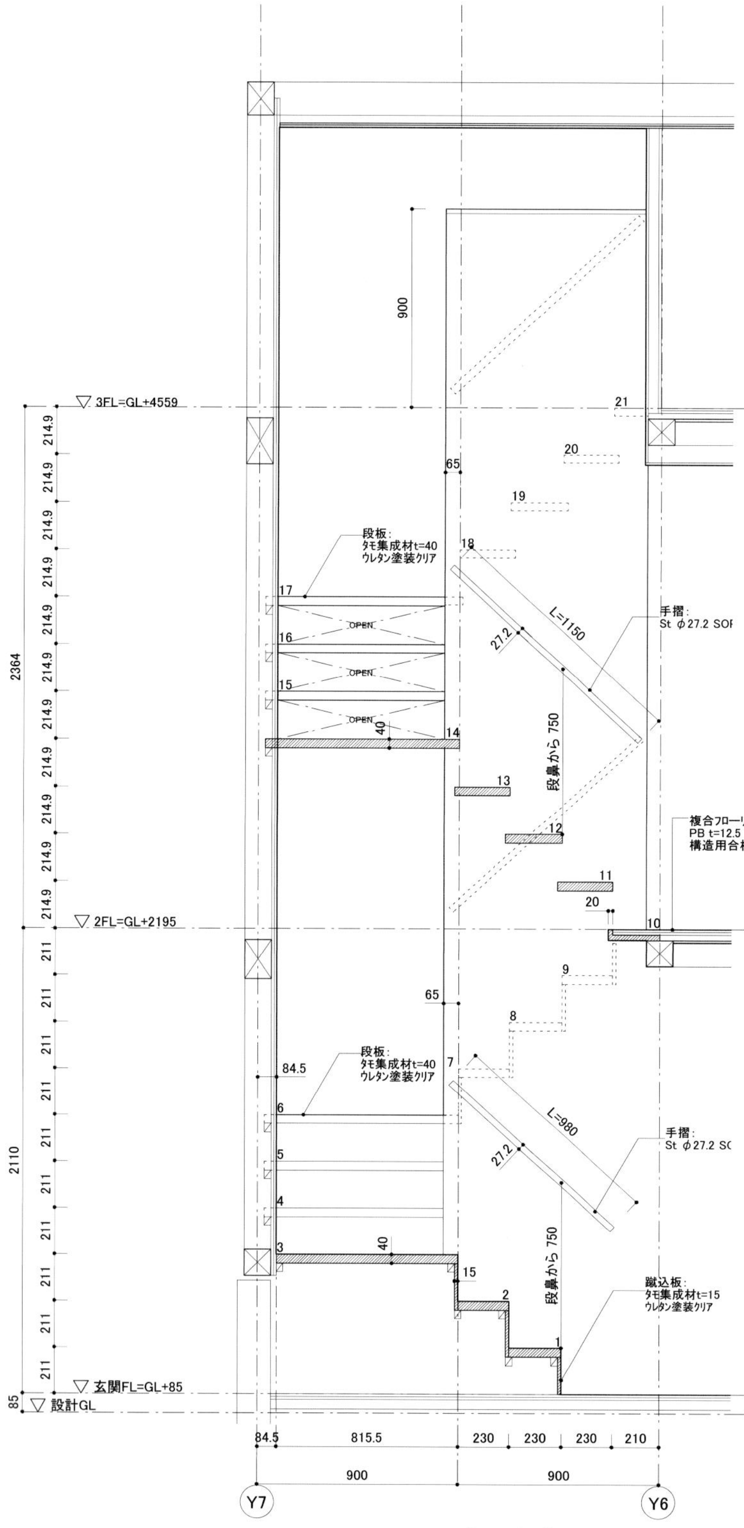

SECTION (S=1/30)

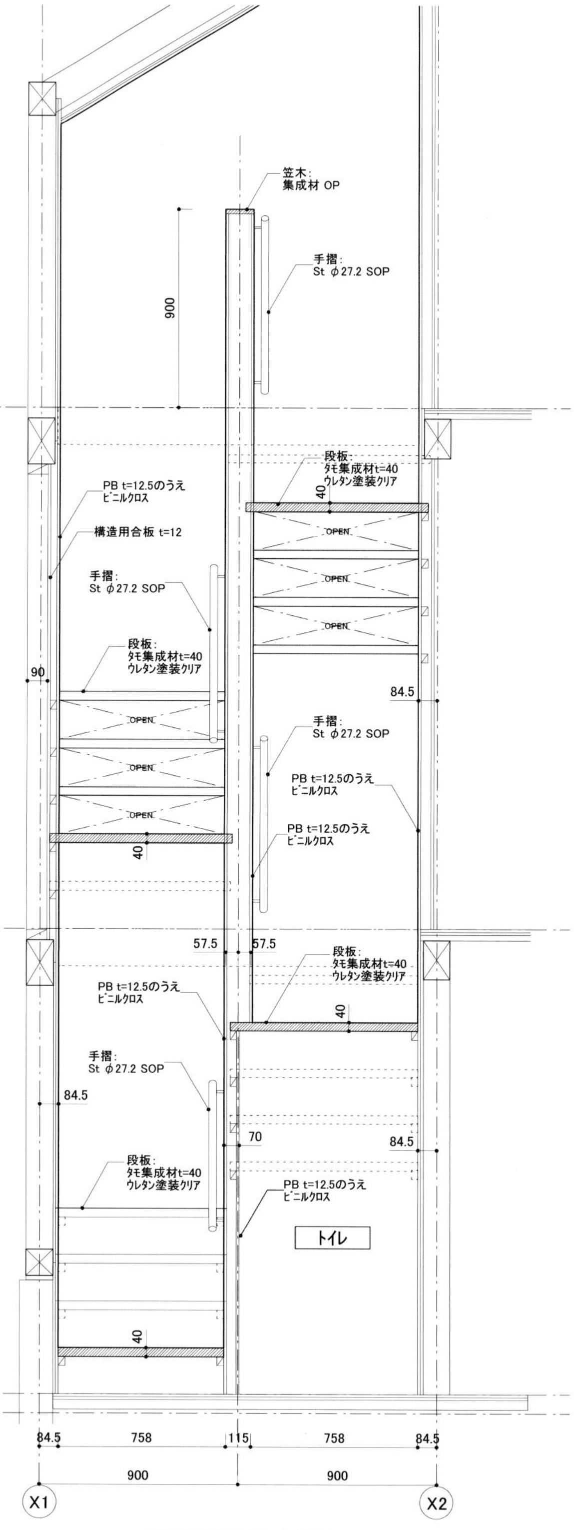

SECTION (S=1/30)

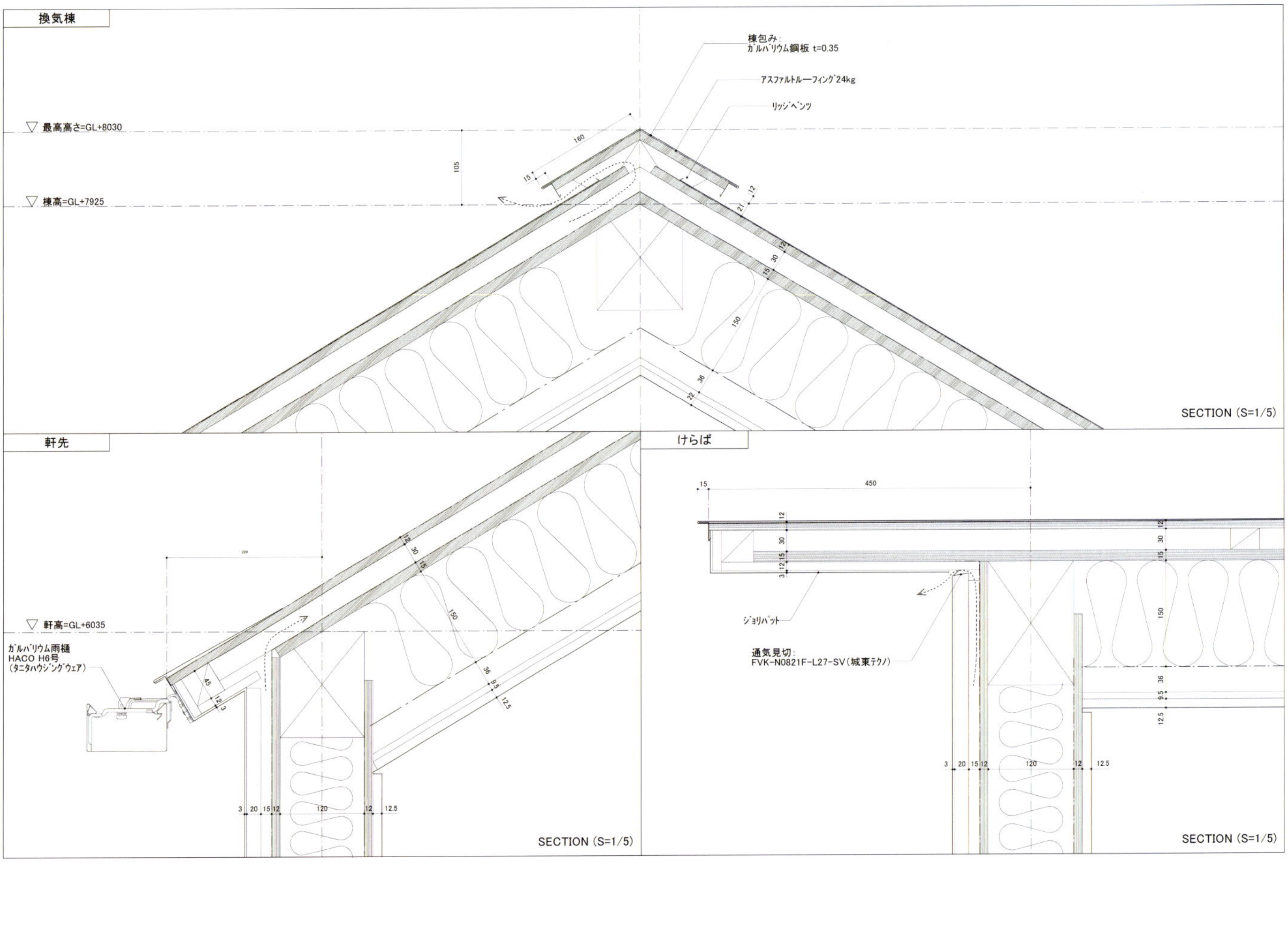
換気棟
棟包み:
ｶﾞﾙﾊﾞﾘｳﾑ鋼板 t=0.35
ｱｽﾌｧﾙﾄﾙｰﾌｨﾝｸﾞ24kg
ﾘｯｼﾞﾍﾞﾝﾂ
最高高さ=GL+8030
棟高=GL+7925
SECTION (S=1/5)
軒先
軒高=GL+6035
ｶﾞﾙﾊﾞﾘｳﾑ雨樋
HACO H6号
(ﾀﾆﾀﾊｳｼﾞﾝｸﾞｳｪｱ)
SECTION (S=1/5)
けらば
ｼﾞｮﾘﾊﾟｯﾄ
通気見切:
FVK-N0821F-L27-SV(城東ﾃｸﾉ)
SECTION (S=1/5)

| **138.99**sq.m. | **Concrete** & **Steel** |

Jubako House

by Meguro Architecture Laboratory

Location _ Tokyo, Japan
Architects in Charge _ Mitsuru Hirai, Sayu Yamaguchi
Site Area _ 126.87m^2
Building Area _ 69.92m^2
Total Floor Area _ 138.99m^2 (1st floor area 68.30m^2, 2nd floor area 42.95m^2, 3rd floor area 27.74m^2)
Structure _ Reinforced Concrete, Galvanized steel sheet on wooden structure
Construction Companies _ Hirano Kensetsu
Structural Design _ Tatsumi Terado Structural Studio
Photography _ Koichi Torimura, Taku Hata

Design of a stacked house 'Jubako house' is based on our concerns on the relationship between the house and the surrounding neighborhood. Jubako: Japanese food stacked boxes, is a set of food containers which can be stacked to sort various kinds of food into each layer. We conceptualized Jubako to create space from a series of the boundary where the perimeter of internal space can be linked with external space. Therefore, we carefully thought about the dimension of the building, choices of material and sizes of openings.

Jubako house is a three-story house for 3 residents. A young couple's room is on the 1st floor and a parent'room is on the 3rd floor. On the 2nd floor, there are living, dining, and kitchen so it is easy for the residents to access from their own floor. We arranged vertical circulation: stairs and elevator, and toilets on the west side of each floor. It is not only beneficial for arranging the plumbing system, but it also prevents the interior space exposing to the sunlight in the daytime. Other utilities such as a bathroom, laundry space, and studio area are on the 1st floor. Structure of Jubako house can be divided into two parts: RC structure on 1st floor and wooden structure on 2nd and 3rd floor covered by galvanized steel sheets. RC structure was designed to be the main structure for the 1st floor and a cantilever parking space. In order to balance weight in structure, we used wooden structure on the 2nd and the 3rd floor. To create opening on the 2nd floor we designed a setback wall with an opening height of 1.2meters behind RC parapet height of 1.8 meters. The height of parapet is higher than the height of the opening; therefore, it can prevent the residents from interfering with the outside. As a result, this opening does not need any blind or curtain, so the perimeter of interior space can be seen continually to the RC parapet. Moreover, this parapet behaves like a reflector of weather conditions where we can see a shadow on sunny day or spots of raindrops on it.

Traditionally, when we discussed 'proportion' of architecture, two issues were usually mentioned. First, the effect of the distance between an observer and a building made a roof or roof decoration object looked smaller than its actual size. In order to make the object look as big as the actual size, it is necessary to draw a larger drawing than its correct proportion on the architectural drawing. Secondly, the portion from the middle of the roof to the top of the roof became blind spot when looking from the ground level. In this case, the height of the roof would be raised up on the drawing. These two proportion issues are based on the human perspective. This is why, when you look at the drawings of temples, roofs particularly are drawn with unbalanced proportion. In the same way, the stacked house 'Jubako house' also has openings where they could not be seen because of the parapet. We also concerned about choices of finishing and rhythm of joints intended to make the building in harmony with the environment.

Nowadays, the concrete issue like 'proportion' has been less concerned than in the past. More conceptual issues such as environment and sociality have become major topics in architectural professions. However, as long as we, architects, create architecture. Such architecture will physically exist throughout time to present the idea behind.

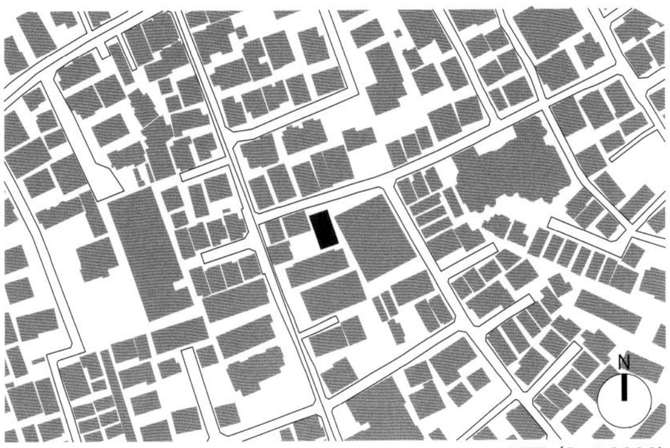

SITE (S=1:3000)

1. Car park
2. Entrance hall and studio space
3. Bedroom
4. Walk-in closet
5. Lavatory
6. Bathroom
7. WC
8. Living and Dining space
9. Kitchen
10. Food storage
11. Balcony
12. Japanese-style room
13. Built-in storage space
14. Garden

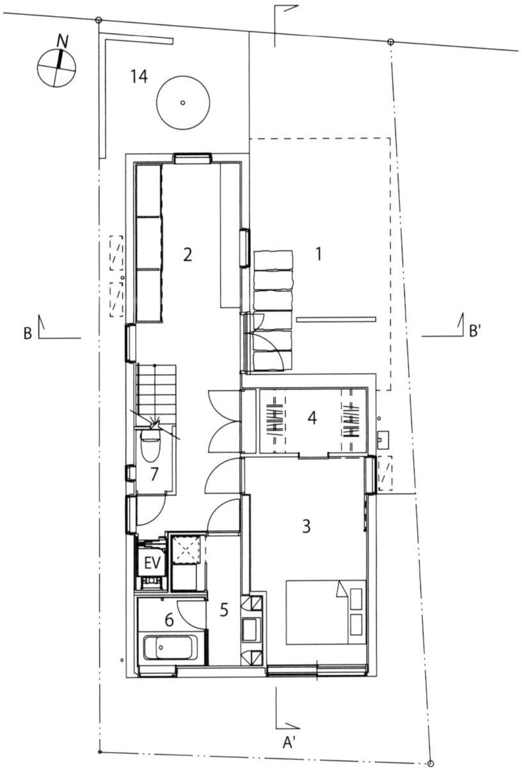

1F PLAN (S=1:150)

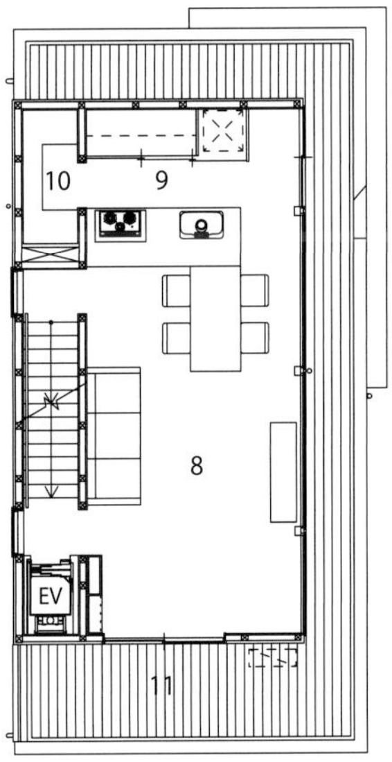

2F PLAN (S=1:150)

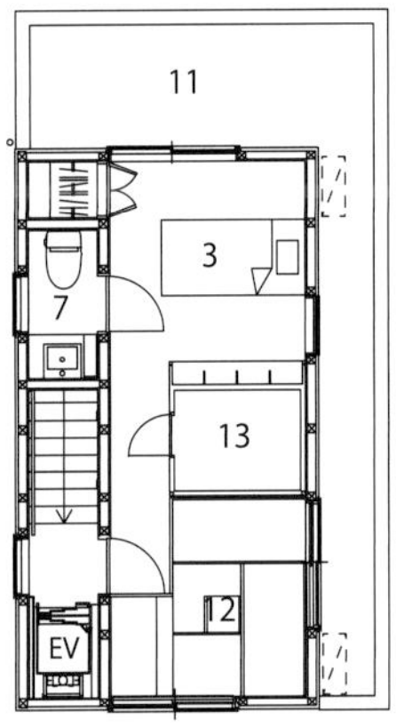

3F PLAN (S=1:150)

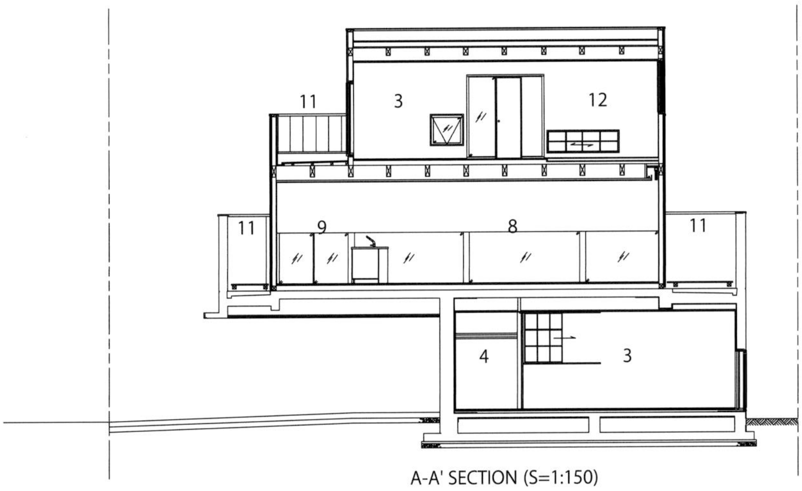

A-A' SECTION (S=1:150)

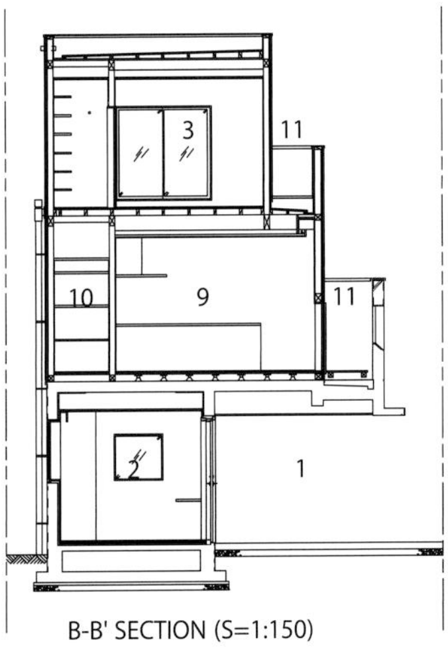

B-B' SECTION (S=1:150)

Isometric diagram of Structural system

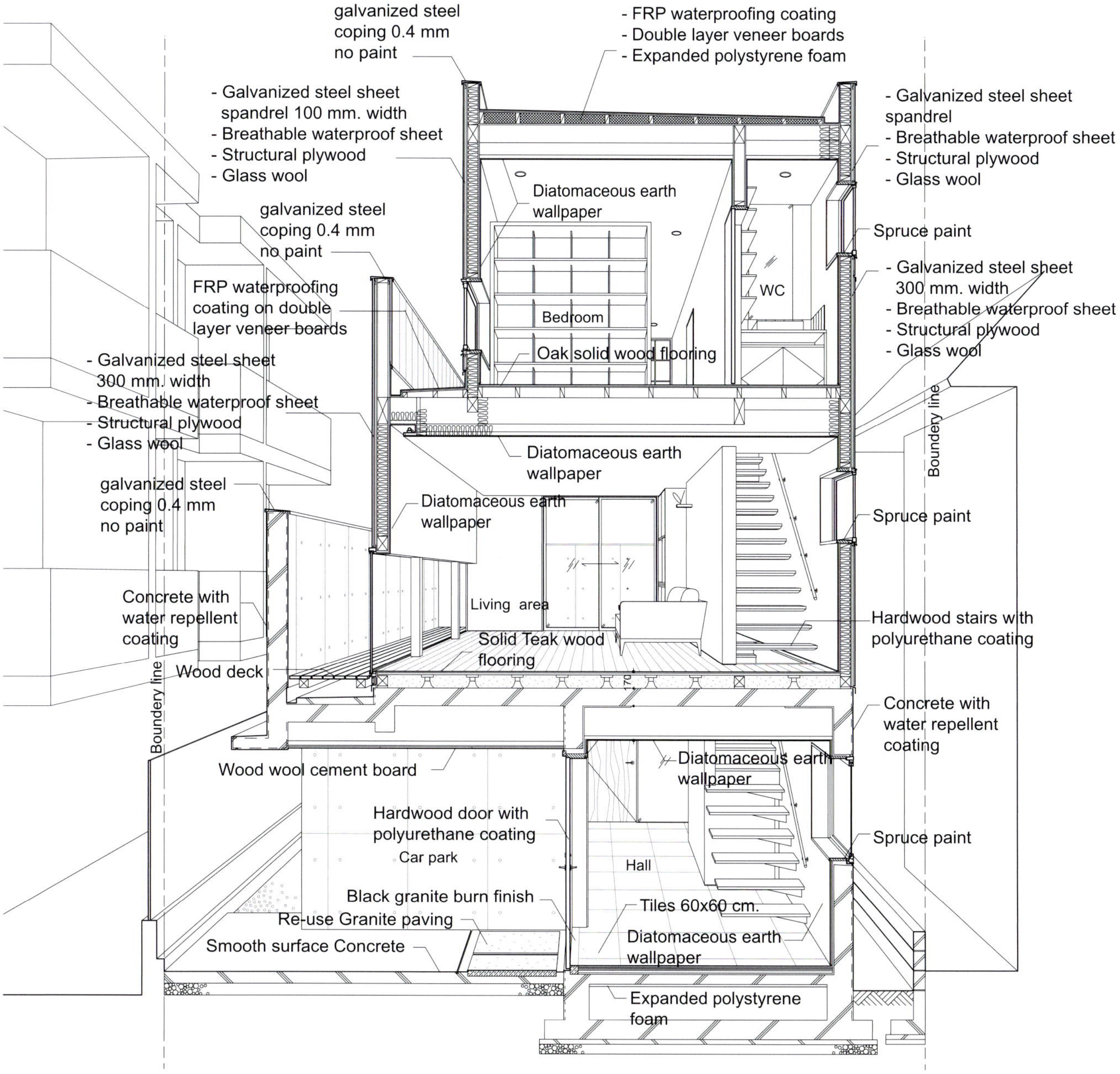
galvanized steel
coping 0.4 mm
no paint
- FRP waterproofing coating
- Double layer veneer boards
- Expanded polystyrene foam
- Galvanized steel sheet
spandrel 100 mm. width
- Breathable waterproof sheet
- Structural plywood
- Glass wool
- Galvanized steel sheet
spandrel
- Breathable waterproof sheet
- Structural plywood
- Glass wool
Diatomaceous earth
wallpaper
galvanized steel
coping 0.4 mm
no paint
Spruce paint
- Galvanized steel sheet
300 mm. width
- Breathable waterproof sheet
- Structural plywood
- Glass wool
FRP waterproofing
coating on double
layer veneer boards
WC
Bedroom
Oak solid wood flooring
- Galvanized steel sheet
300 mm. width
- Breathable waterproof sheet
- Structural plywood
- Glass wool
Boundery line
Diatomaceous earth
wallpaper
galvanized steel
coping 0.4 mm
no paint
Diatomaceous earth
wallpaper
Spruce paint
Concrete with
water repellent
coating
Living area
Hardwood stairs with
polyurethane coating
Solid Teak wood
flooring
Wood deck
Boundery line
Concrete with
water repellent
coating
Wood wool cement board
Diatomaceous earth
wallpaper
Hardwood door with
polyurethane coating
Car park
Spruce paint
Hall
Black granite burn finish
Tiles 60x60 cm.
Re-use Granite paving
Diatomaceous earth
wallpaper
Smooth surface Concrete
Expanded polystyrene
foam

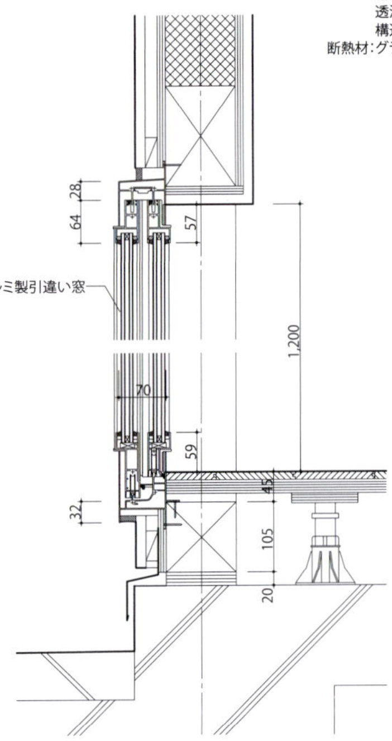

引違い部断面詳細図　S=1:10

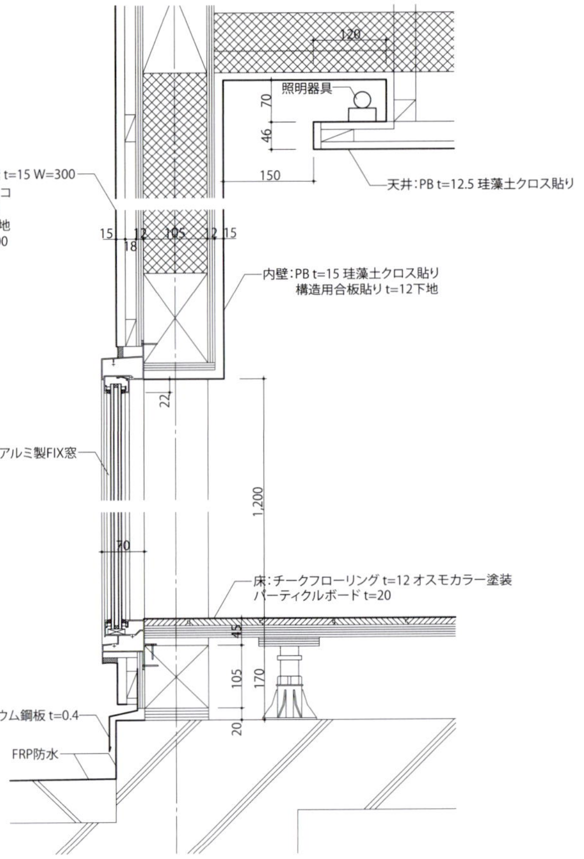

FIX部断面詳細図　S=1:10

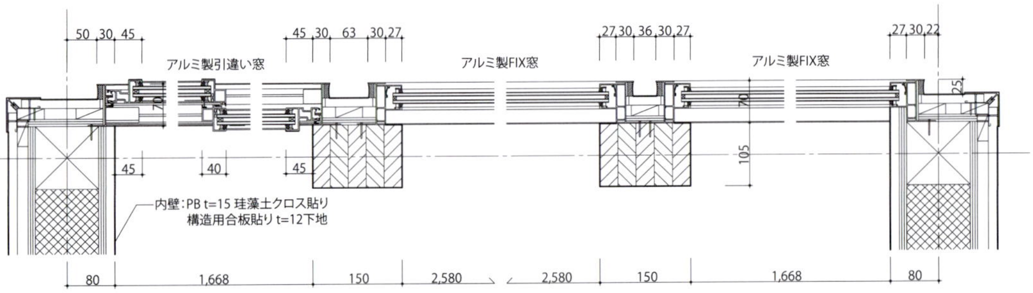

2階開口部廻り平面詳細図　S=1:10

| 143.95sq.m. | Concrete |

PEAK

by APOLLO Architects & Associates

Location _ Sumida ward Tokyo
Site Area_ 89.30m²
Building Area _61.11m²
Total Floor Area _ 143.95m²
Structure _ RC
Scale _ 3F
Structure engineer _ Masaki structure laboratory, Kenta Masaki
Facility engineers _ Naoki Matsumoto
Constructor_ Shinosaki Building contractor's office
Photography _ Masao Nishikawa

This home is located in the old town area of eastern Tokyo, where the atmosphere of an industrial district still lingers. The client, who works at a large advertising agency, and his wife purchased the building site in an area where they had lived in for some years. Their primary request with regard to the design was that the house has a courtyard.

The three-story structure has reinforced concrete walls on the three sides facing neighboring homes, with open lines of sight only on the side facing the road. To prevent passersby from seeing inside, one-way mirror glass covers the large openings on the street side. On the southeast side, a courtyard topped with a void brings light and air into the home without loss of privacy, creating a resort-like atmosphere despite the urban location.

The piloti-style garage on the first floor is separated from the covered entryway by a glass wall and steel door for enhanced security. The first floor includes a bathroom with an adjacent miniature garden of the type often found in hotels and the client's study. The study is floored with tatami mats and can be used as a guest room as well as for working at the Japanese-style desk. On the second floor, a large open-plan living, dining, and kitchen area faces the courtyard, bringing the pleasures of outdoor living into the urban setting. The low ceiling in the kitchen and dining areas contrasts with the two-story void above the living room and courtyard, which interact dynamically as the light from the skylight plays over them. On the third floor, the master and children's bedrooms sit side by side. A bridge, which is accessible from the bedrooms, offers a sweeping view over the interior and exterior scenery.

This small urban home represents a contemporary take on Japan's machiya-style townhouse—here arranged vertically rather than horizontally, as is traditional—that offers a variety of places for the residents to relax as they move through space.

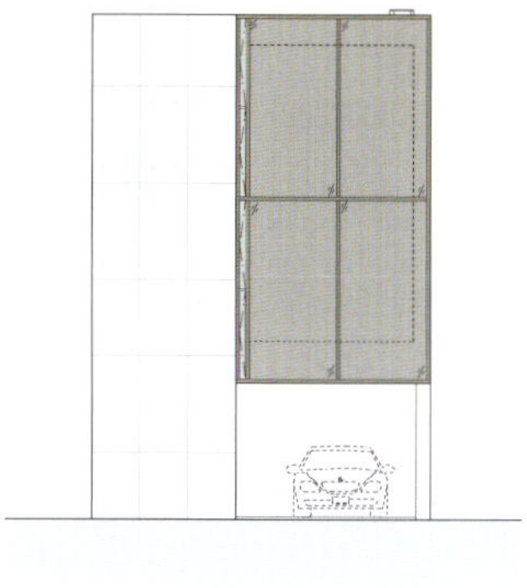

East Elevation

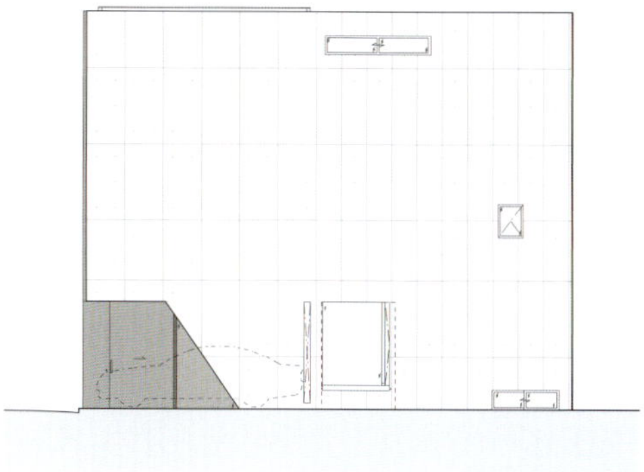

North Elevation

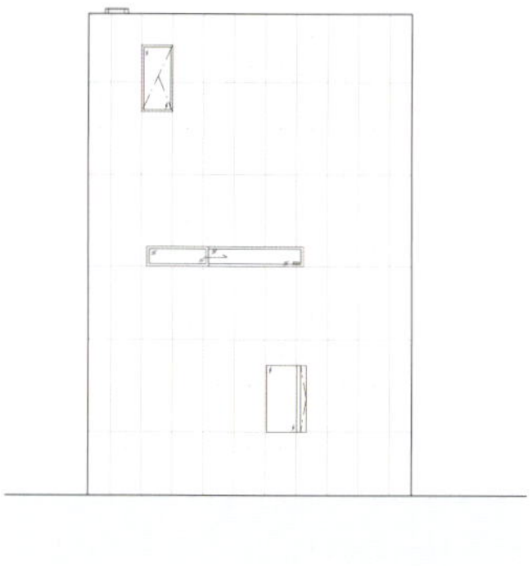

West Elevation

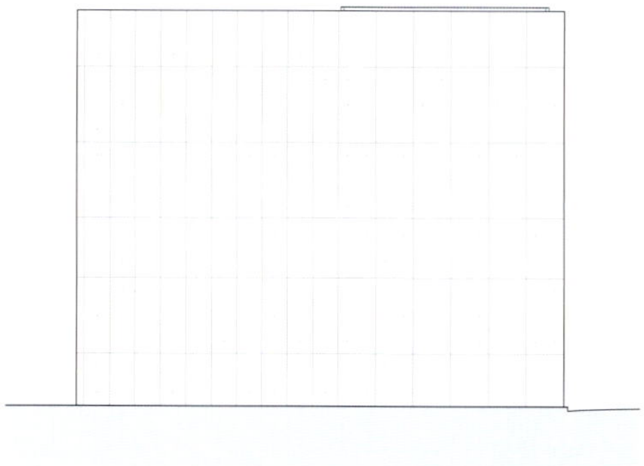

South Elevation

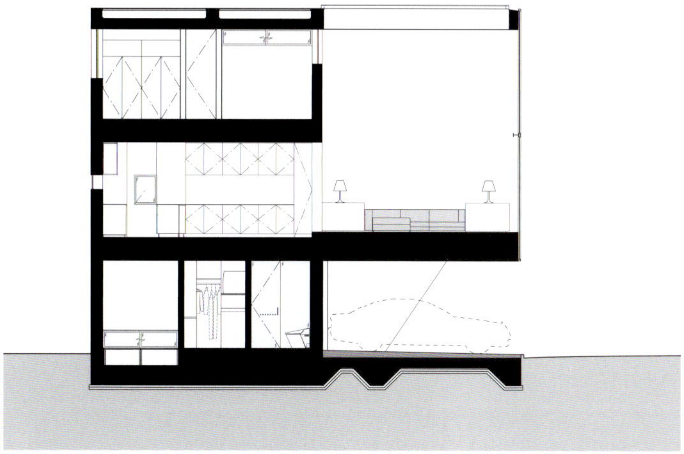

Section A-A'

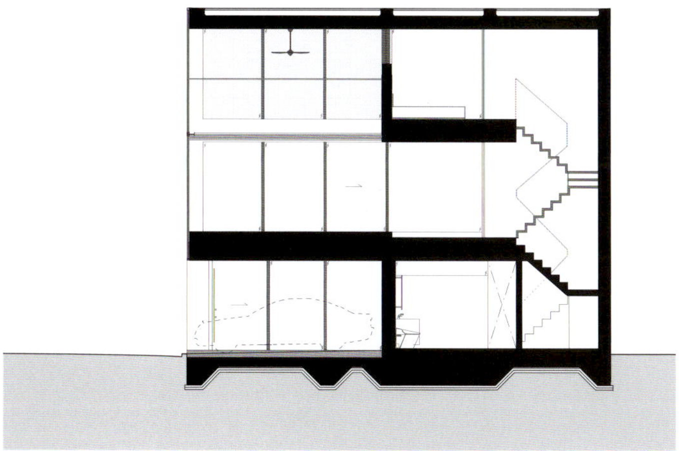

Section B-B'

Section C-C'

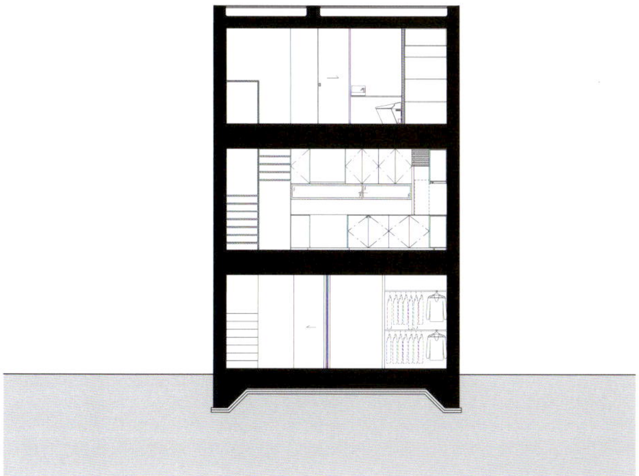

Section D-D'

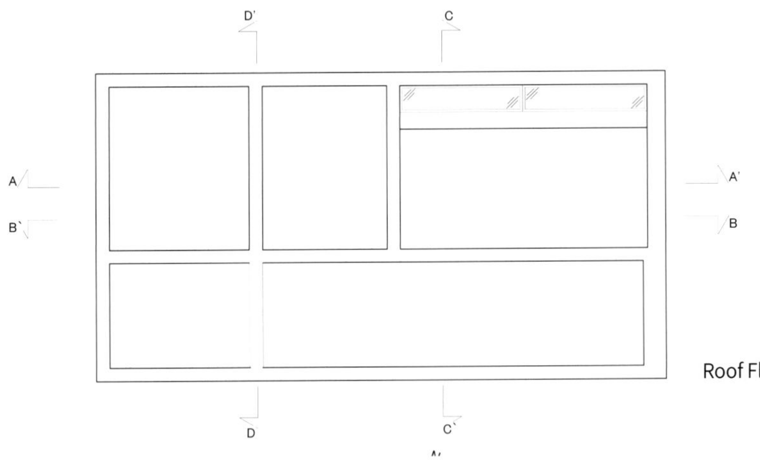

Roof Floor

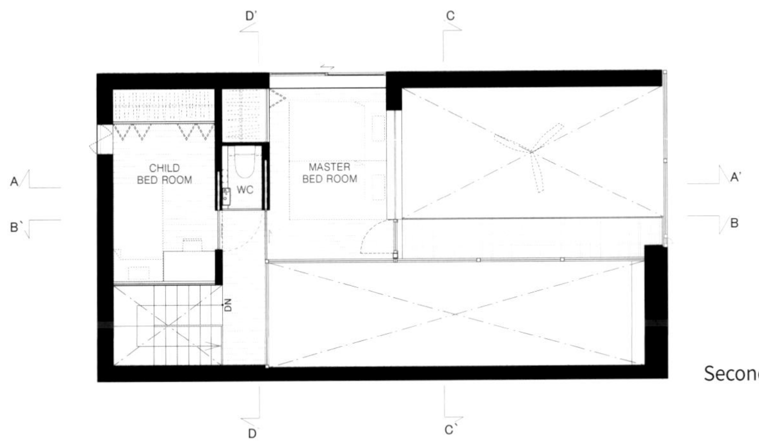

Second Floor

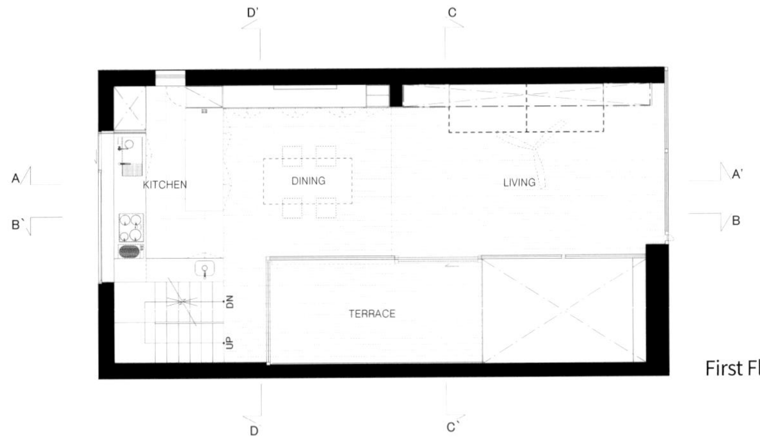

First Floor

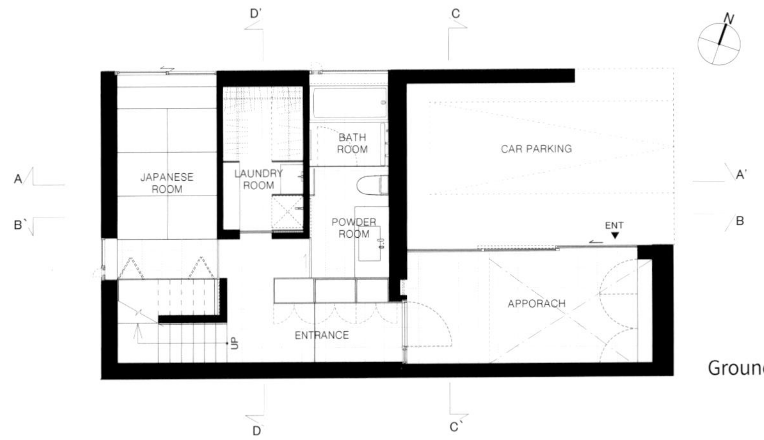

Ground Floor

| **152.22**sq.m. | **Timber** |

House in Nikko

by TAKiBI

Location _ Tochigi in Japan
Total Floor Area_ 152.22m²
Photography _ Koji Fujii (Nacasa&Partners)

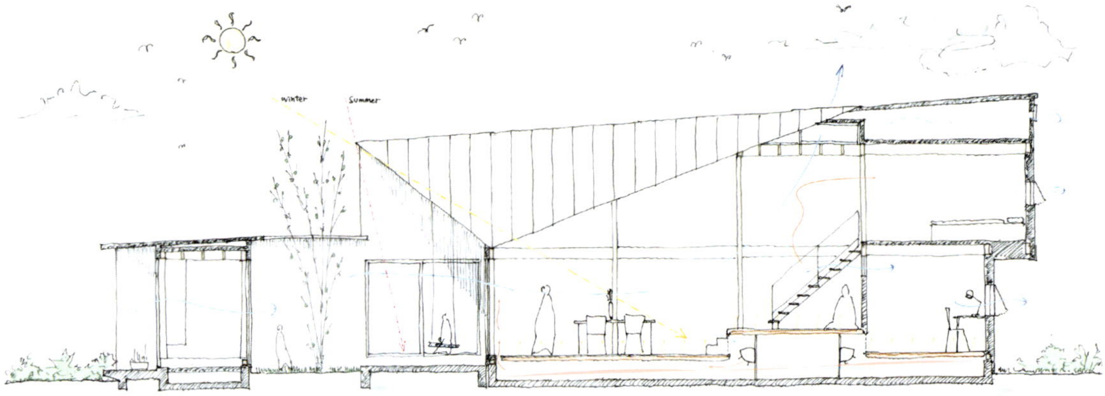

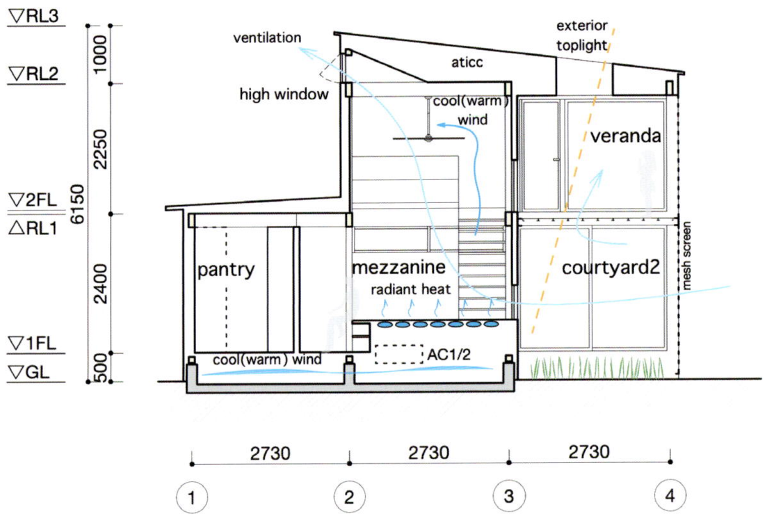

Short Sides Section (Summer Diagram)

We arranged alternately two courtyards and rooms under the diagonally big roof and horizontal roof. The form of diagonally plan and roof are designed based on the orientation and angle of solar radiation, prevailing wind, surrounding buildings. As a result, we created modulated space which is composed of LDK in the stairwell and low service area. This house adopted a complete outside heat insulating method including the foundation. We adopted an air-conditioning system named "air aqua." This system is composed of an air conditioner installed under the mezzanine floor and water pack set up under the LDK floor. This system's advantages are high efficiency and energy saving and low-cost. NIKKO is hot in summer and cold in winter, but we created an expansive house which has a high-performance thermal environment by integrating passive environmental control method and active method. An assembly room next to the entrance is using for a meeting of the neighborhood. We are aiming for life to be opened to the city by letting a part of the residence have a public function. As a result of being conscious of that the entire residence space will be connected to the ground in order to support such activities, it became a flat and linear style spatial structure. In addition, the margin between the building and the front road is regarded as an editable area according to the needs of the homeowner and the street, and it is under consideration to extend exterior and multipurpose huts. we hope that the birth of this building will create a positive impact on the community.

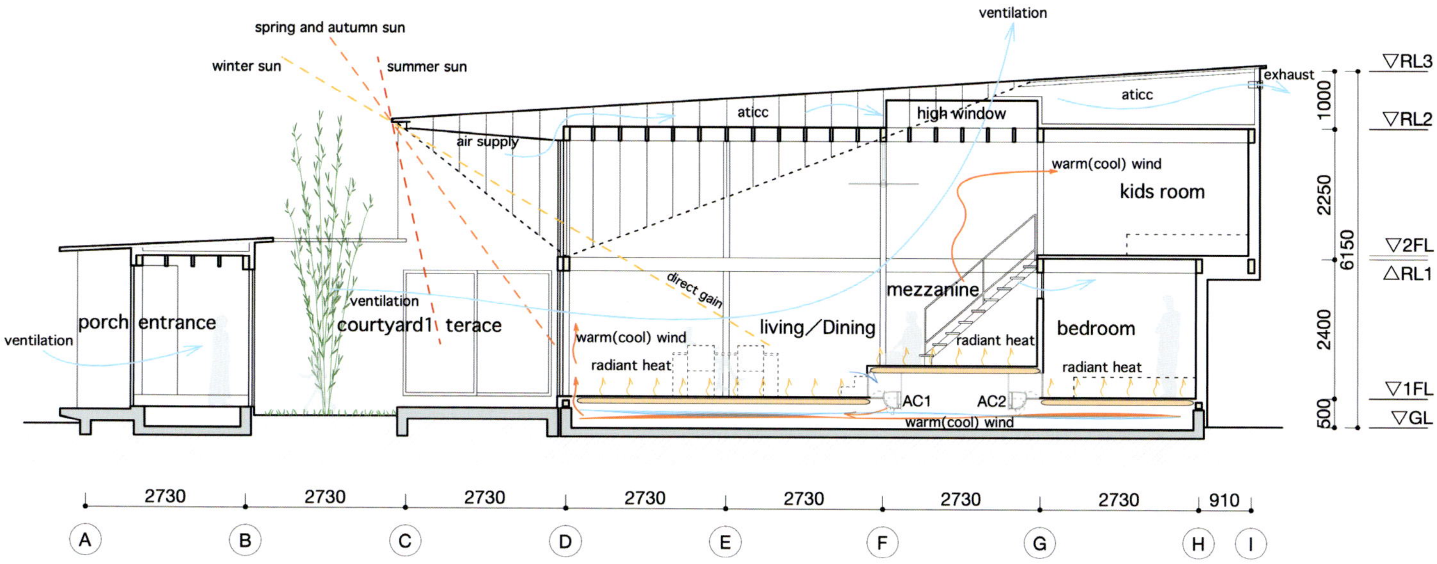

Long Sides Section

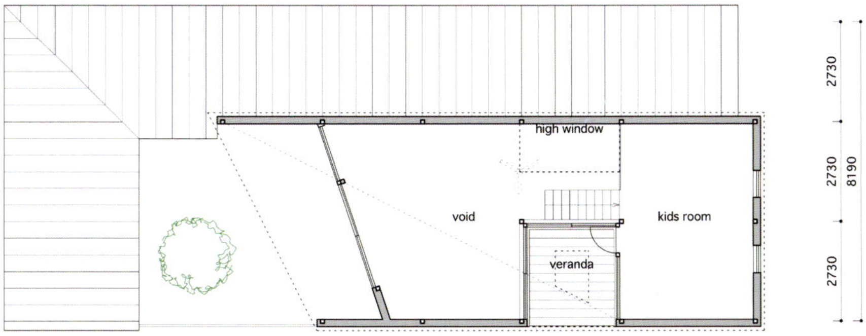

First Floor

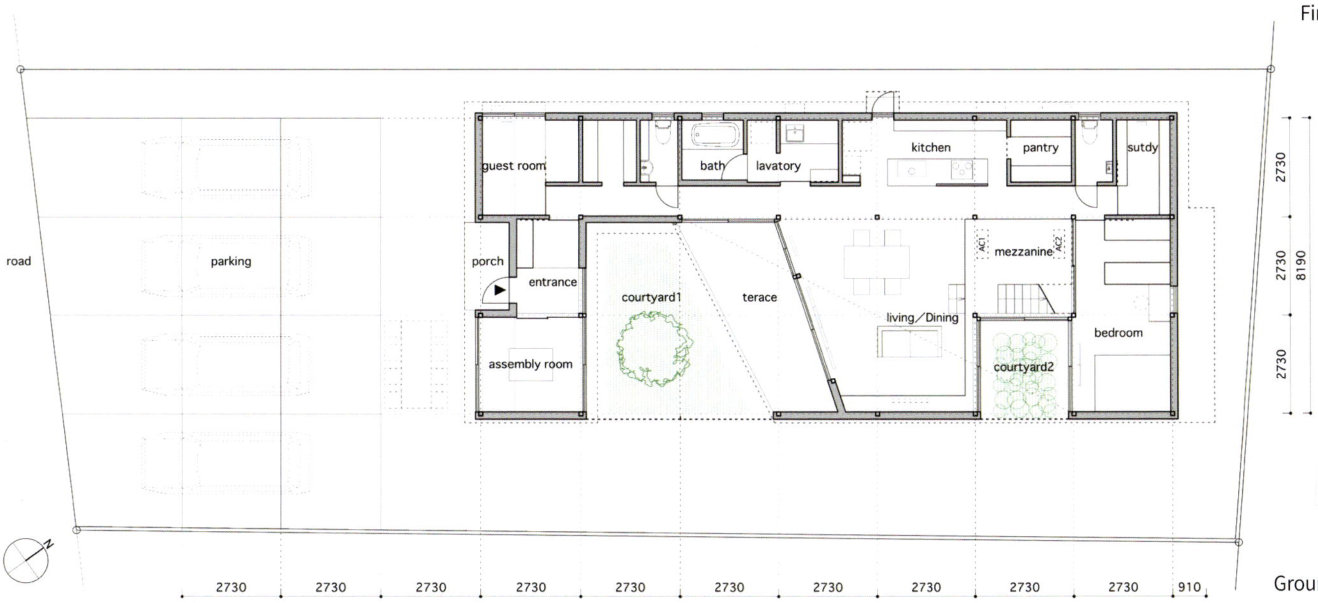

Ground Floor

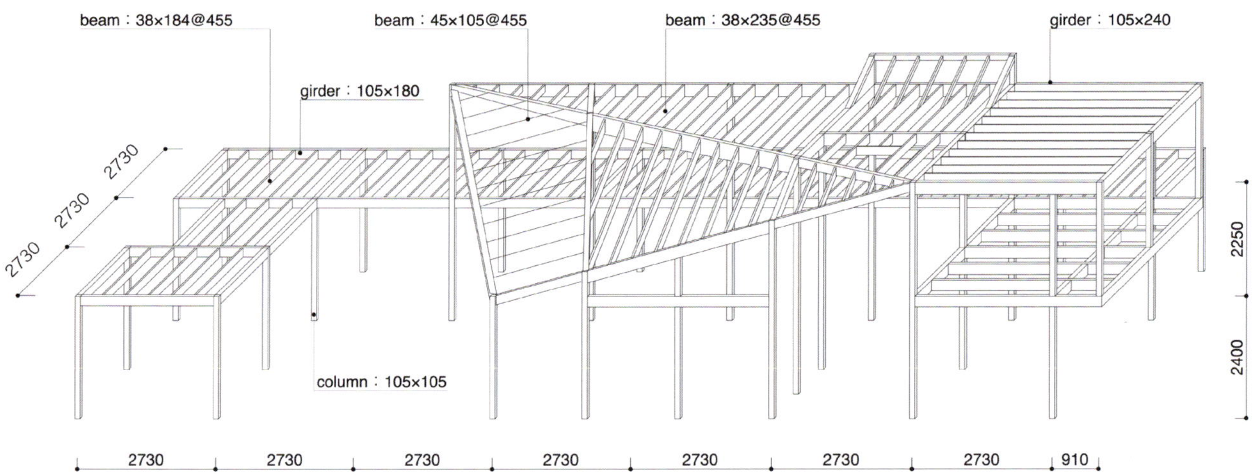
beam : 38×184@455
beam : 45×105@455
beam : 38×235@455
girder : 105×240
girder : 105×180
column : 105×105
2730
2730
2730
2250
2400
2730
2730
2730
2730
2730
2730
2730
910

軒裏の**給気口**から小屋裏へ取入れた空気と外壁通気層の空は壁頂部の換気ベントキャッから排出し夏季の**輻射熱を制**しています。

大屋根の**斜めの折り下げ**は、住まいの中心に**大きな空間**を確保しながらも、**冷暖房負荷低減**のために**開口面積や気積は抑えたい**という建主の要求に対する解答にもなっています。また、平屋部分や2階の**階高**を、この吹き抜け空間と対比的に**最小限に抑える**ことで**空間的なメリハリ**を生み出しています。

落葉樹であるヒメシャラは将来的に5mほどに成長し**夏期日射制御**の機能を担います。

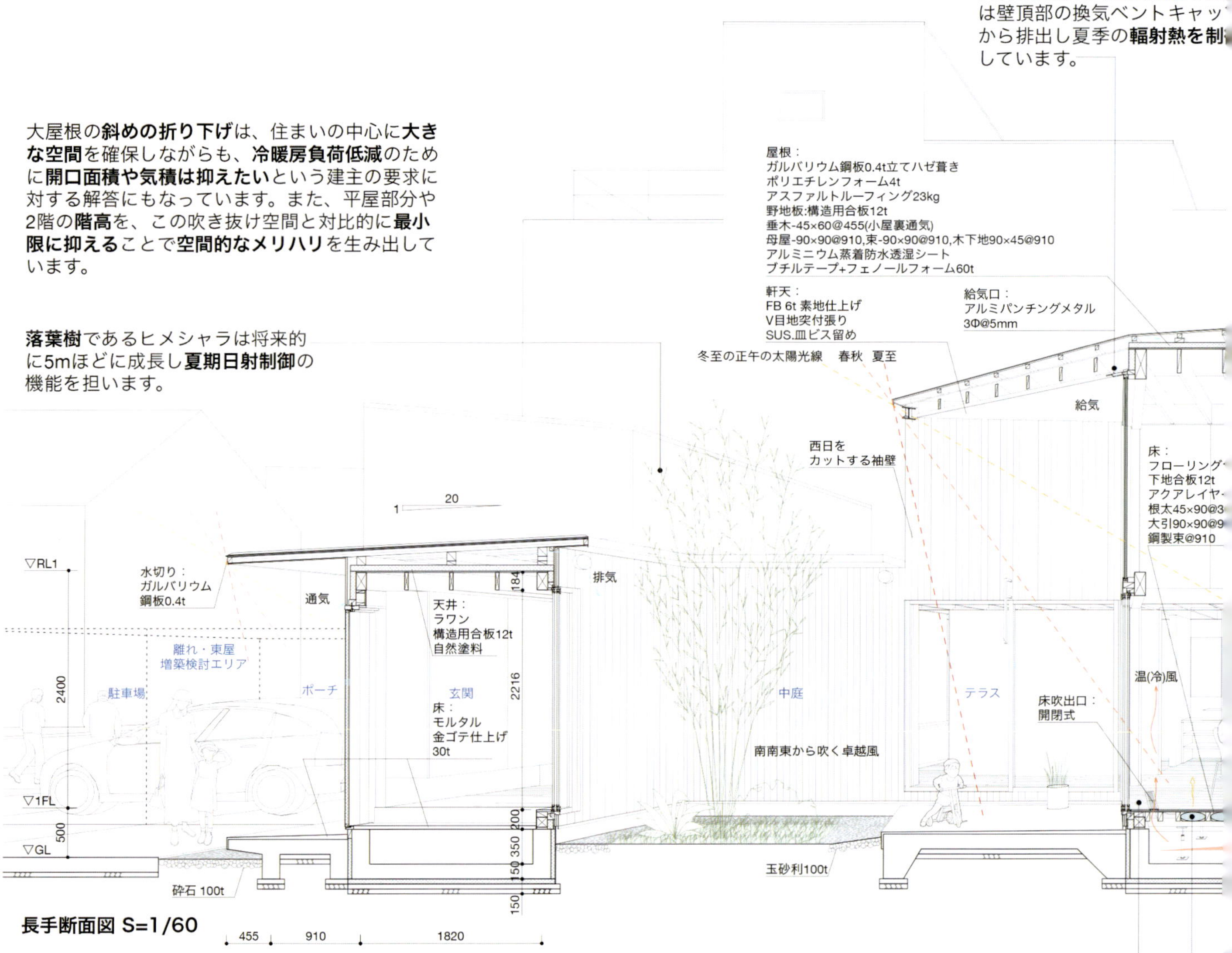

長手断面図 S=1/60

開口部には**アルゴンガス**を封入した**Low-eペアガラス**を**アルミ樹脂複合枠**にはめ込んだ高性能なサッシを採用しています。冬季の日中は南面からの**ダイレクトゲインを取得**できるため暖房の運転は不要ですが、夜間や曇天時に窓面からの冷輻射や**コールドドラフトを軽減**させるために、掃き出し窓の足元に**空調吹出口**を配置しました。

屋根・外壁・基礎共に構造軸組の**外側に発泡断熱パネル**を張った**完全な外断熱構法**としています。外壁仕上げと断熱パネルの間には**アルミニウム蒸着の透湿防水シート**を張り、**小屋裏へ連続する通気層**を設けています。**南面軒裏の通気口**から**小屋裏**へ取り入れた空気と**外壁通気層**の空気は、北壁頂部の換気ベントキャップから排出し**夏季の輻射熱を制御**しています。

シーリングファンにより夏期は床下空調の冷気を上方へ引上げ冬期は暖気を下方へ押し下げるように回転することで、2層分の吹抜け室内空間上下の**温度ムラの解消**を促します。

2010年に東京大学で実施された**実測実験**において、**屋根窓の有効性**が実証されました。特に周囲に建物が立ち並ぶような住宅地では地表よりも屋根面における風の流れが強いことや、壁面の開口部から風が抜けにくいため、**自然通風促進**のための非常に有効な手法です。

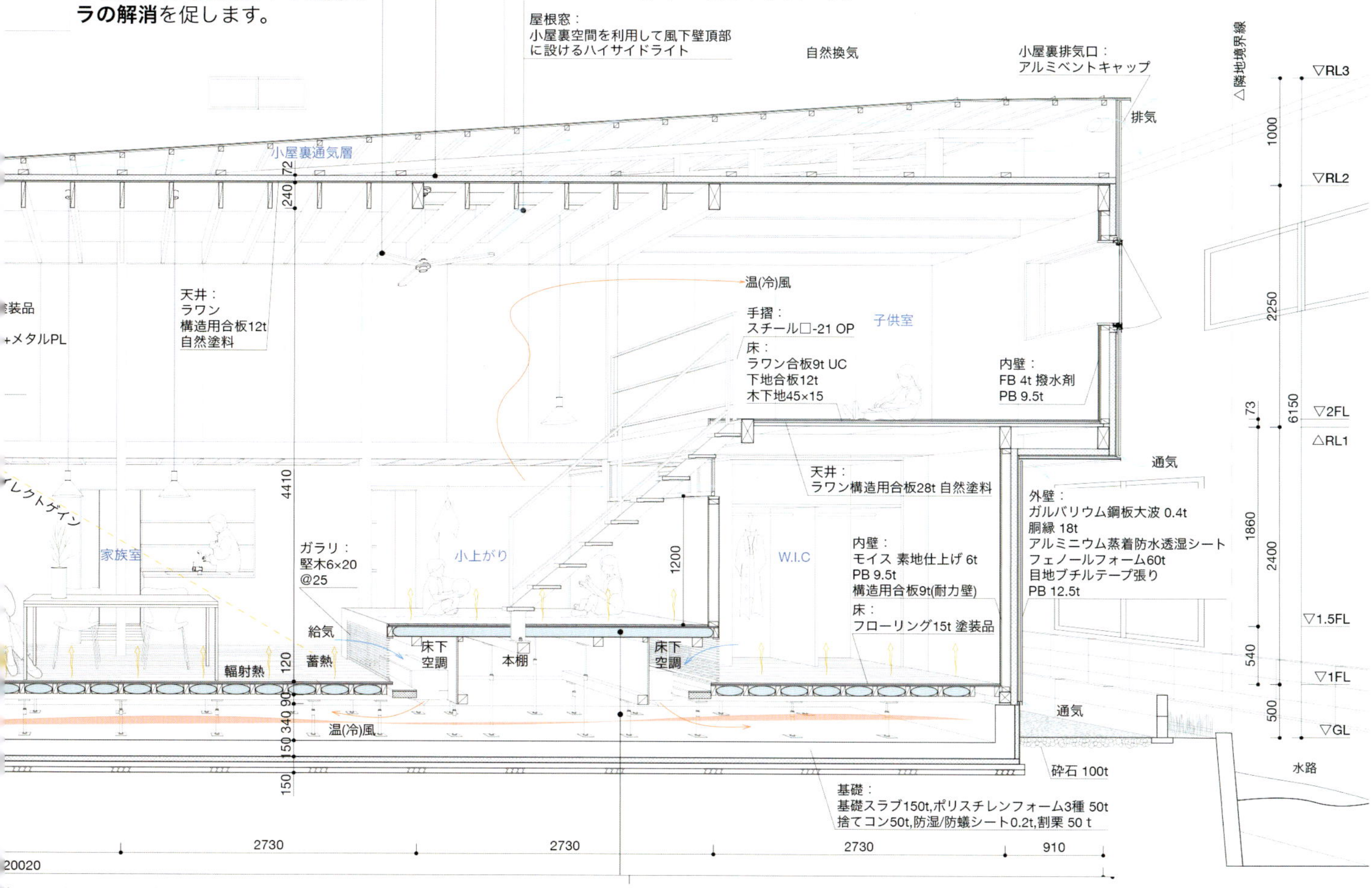

高い比熱と温度差によって**自然対流する「水」**の物性を利用した**蓄熱式床暖房「アクアレイヤーシステム」**と小上がり床下空間を活用した床下空調(家庭用)を組み合わせた**省エネルギーでローコストな空調システム**を採用しています。床下に空調空気を吹き込み、**水袋「アクアセル」と基礎コンクリートに蓄熱**させます。空調空気は開口部側の床面や壁や棚の巾木に設けた**開閉式吹出口**から室内に吹き出し、再び空調器へと還流させます。**暖房と冷房が可能**な点と**輻射と対流の空調を併用**できる点が特徴です。**パッシブな建築的な環境制御**の手法と**アクティブ**な方法を組み合わせることによって、夏は暑く冬は冷え込みの厳しい日光地域においても、**高性能かつ開放的な住まい**を実現しています。

| 157.77sq.m. | Wood |

Gable Roof House

by Alphaville

Location _ Kyoto, Japan
Site Area _ 228.92m^2
Building Area _ 104.37m^2
Total Floor Area _ 157.77m^2
General Contractor _ Kawana Kogyo Co., Ltd. - Takeshi Hara
Structure System _ Wood
Photography _ Yasutake Kondo

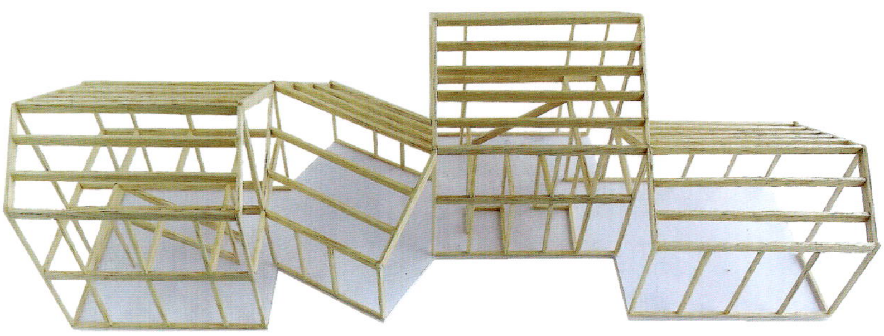

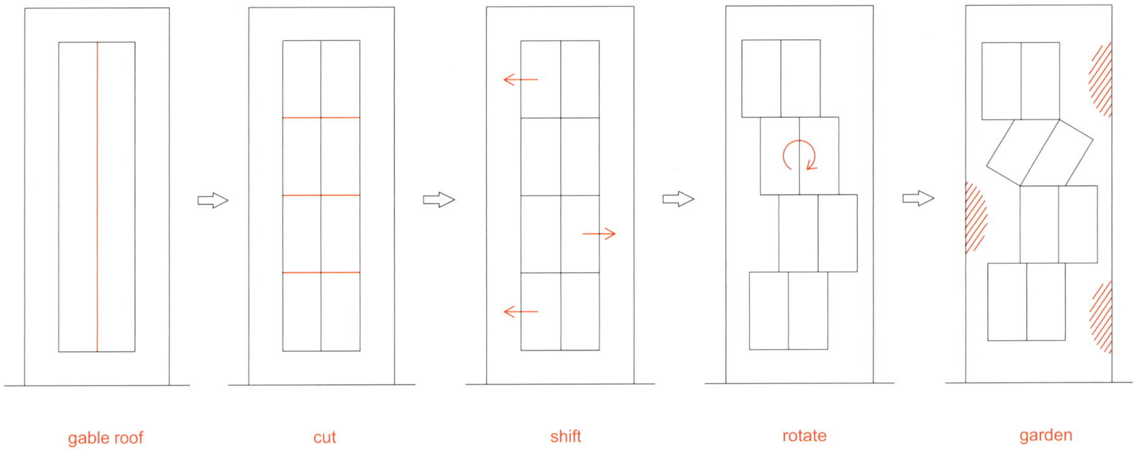

We set out to interpret traditional gabled roofs, the cornerstone of Kyoto's townscape ordinance, within the framework of modern dwellings. As Laugier's "Primitive Hut", gabled roofs hold a sort of fundamental appeal. However, we felt there was a problem in their strong axiality and the fact that sunlight in the house is unevenly distributed. Additionally, we also saw an issue in dealing with the gable's outward thrust. As such, we took a long gabled roof spanning a deep plot of land and divided it into four parts. We shifted the position of the four square houses, starting from the one closest to the road, rotated them and changed their heights. By doing so we created a sheltered space between buildings while weakening the house's axis line. The center of the house is a one-story kitchen-dining area angled at 30 degrees. From this area, one can look into each of the rooms on the first and second floors of the two adjoining wings. A surprising sense of open space is generated by the discontinuous sequence of rooms with varying roof heights and tilts which are built from a combination of twisted single-and double-layered volumes. The house uses a wooden panel structure, but we made it possible to place a large transparent opening in the garden-facing wall by having the roof beam of the single-story structure to act as a brace. In this way, by bringing together different constructional components at boundary divisions and squaring the roof's joist with the gradient, we have changed the historically meaningful gable shape into a structure which does not generate thrust. The shifting and layering of the gabled roof created a space where sunlight-filled active areas overlap with secluded areas for curling up with a book. This project attempted to create deep relationships between each room; between the rooms and the garden; and between both of these spaces and the street.

Site Plan

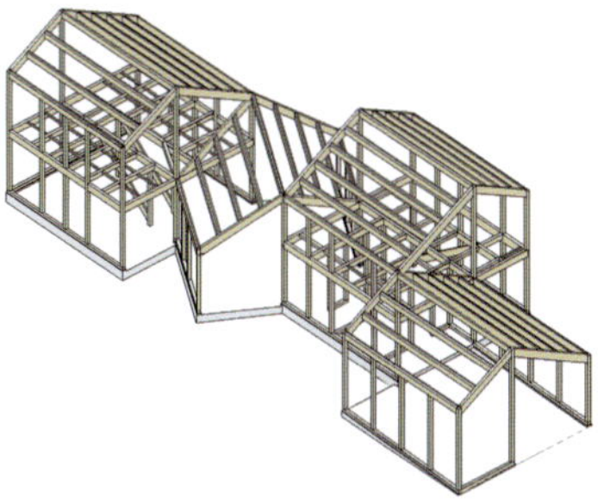
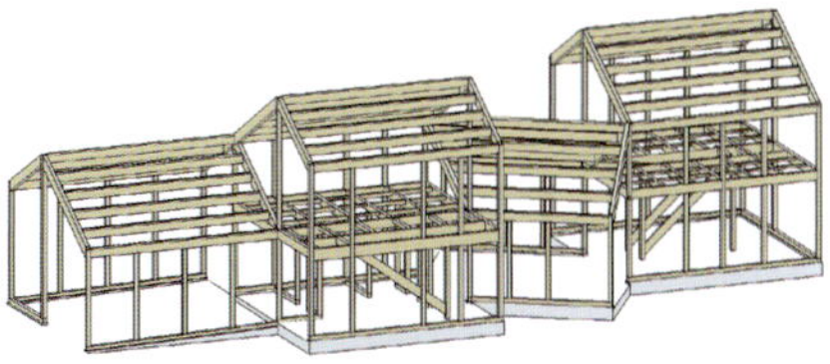
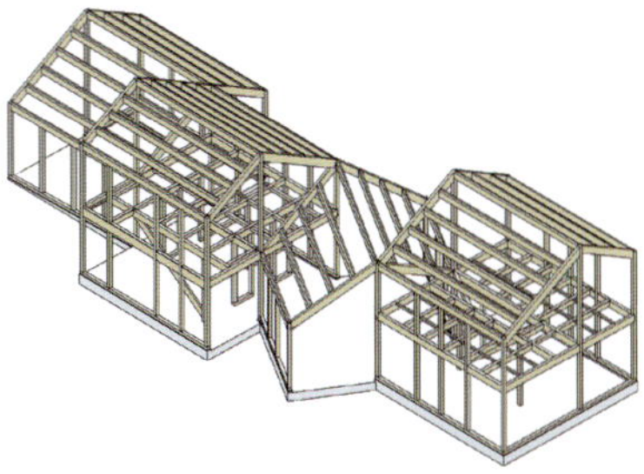

South Elevation

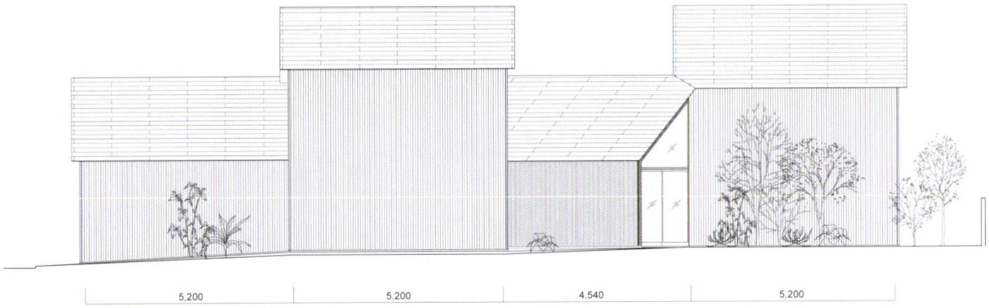

East Elevation

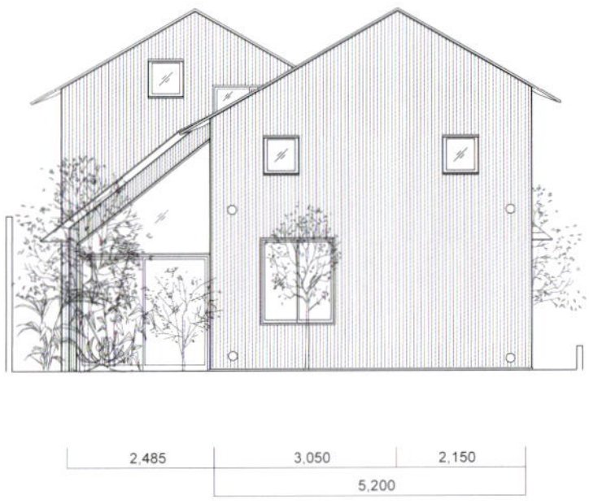

North Elevation

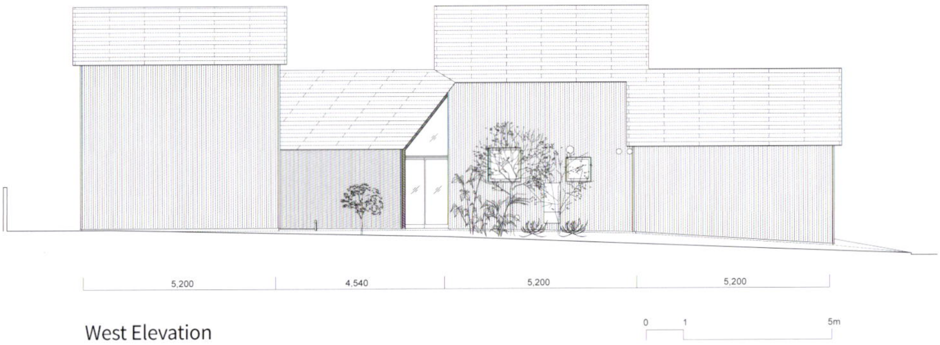

West Elevation

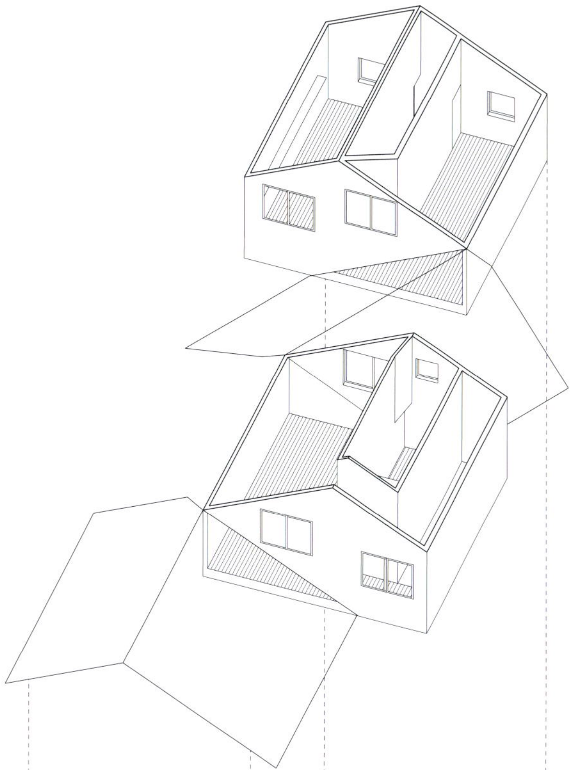

2F

1F

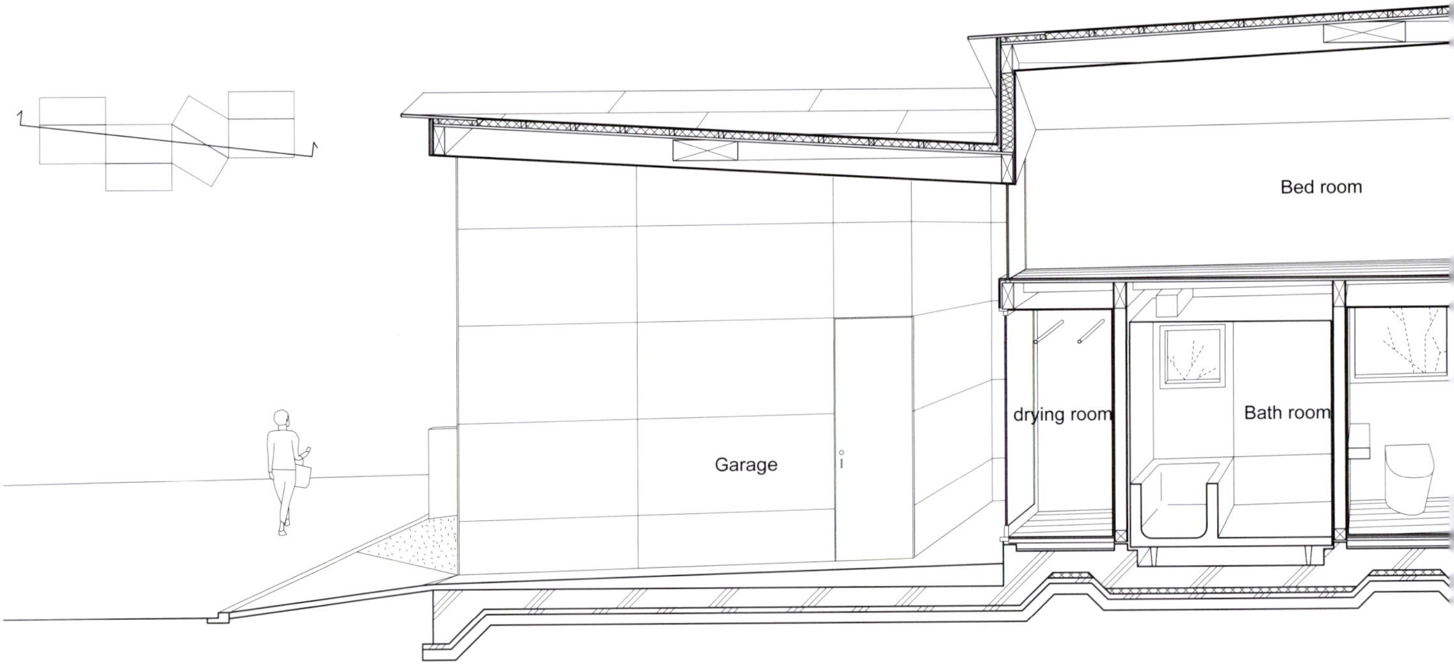
Bed room
drying room
Bath room
Garage

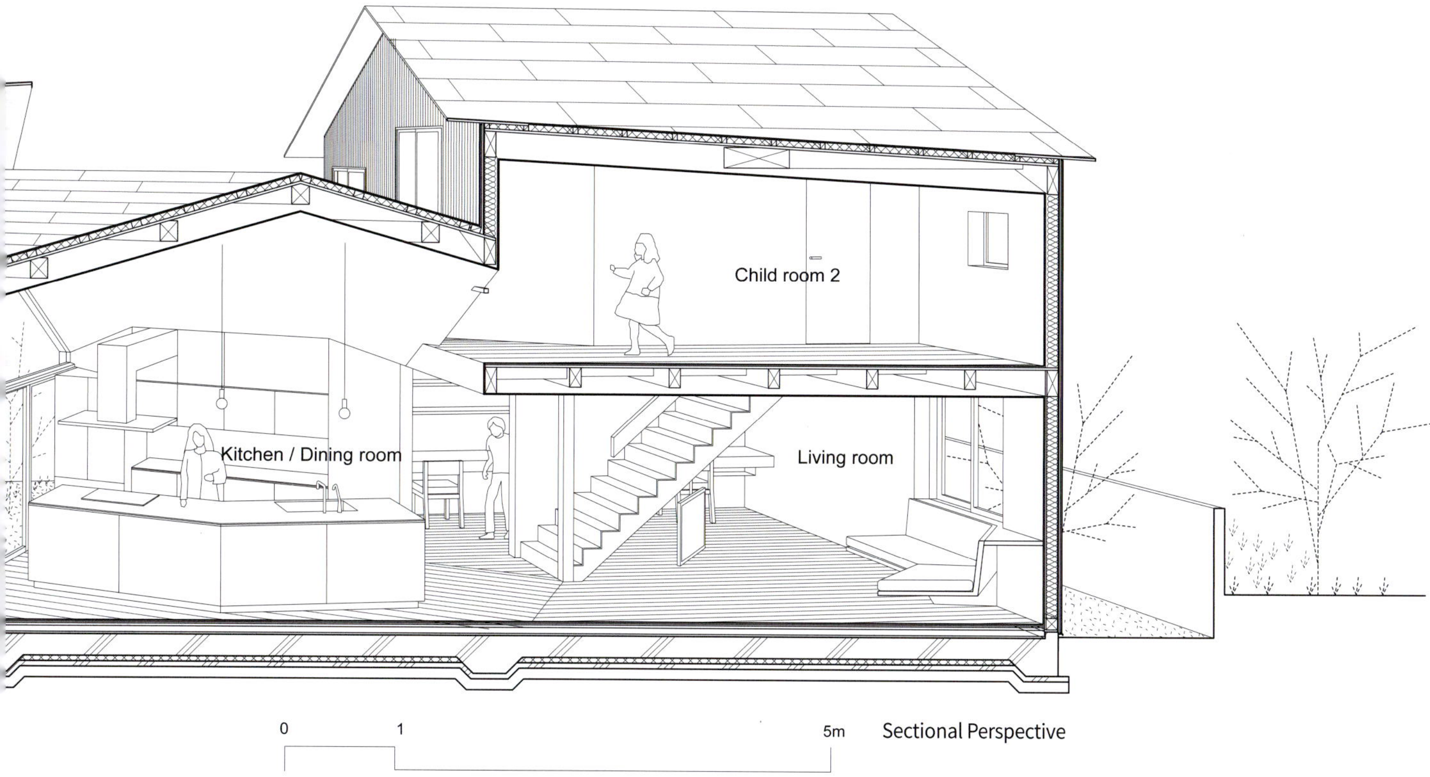

Sectional Perspective

a b c d e f g h i
0 1 5m
5,200
a
2,150 3,050
b
f
2,150 3,050 2,150
c
2,150 3,050
g
3,050 1,250 900
d
2,150 900 2,150
h
5,200
e
5,200
i
Sections

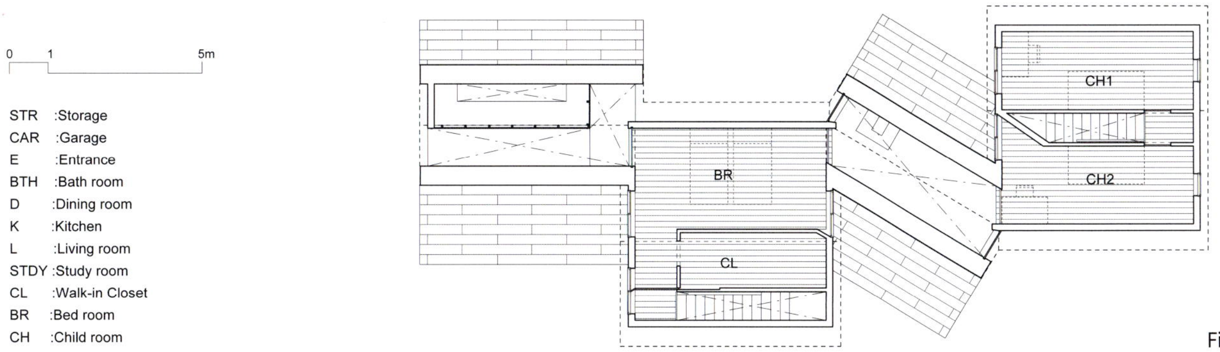

First Floor

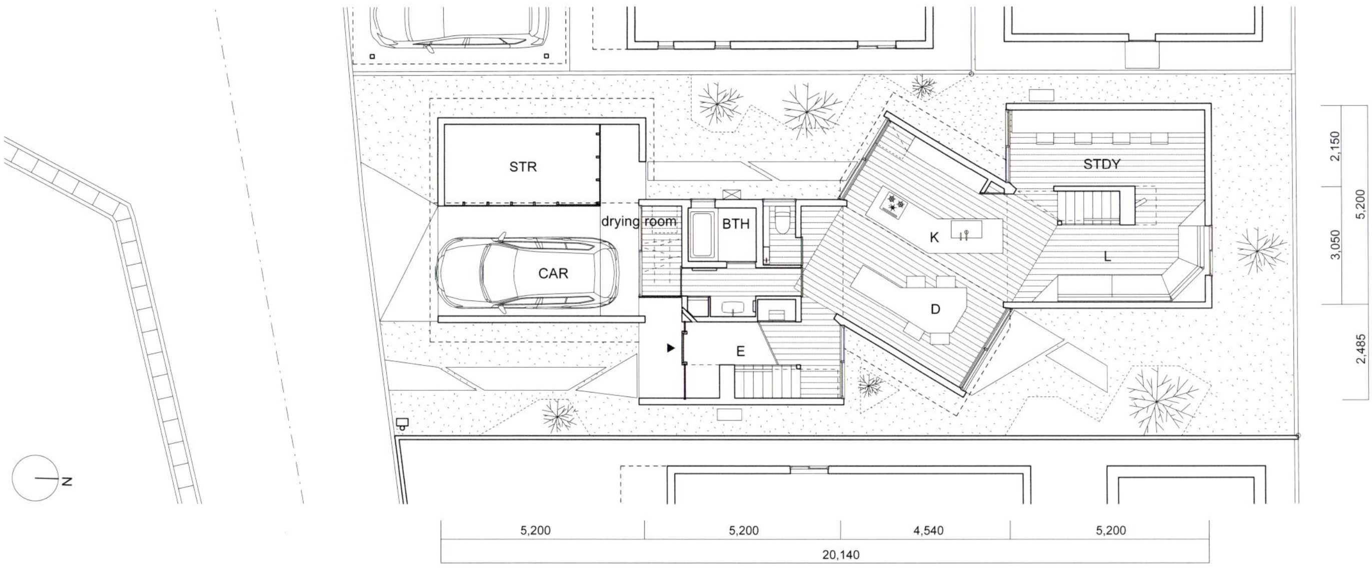

Ground Floor

| 162sq.m. | Steel & Brick |

Folded House

by Proctor & Shaw

Location _ London Borough of Wandsworth, UK
Total Floor Area _ 162m²
Engineering _ DFRYER Consulting Engineering
Photography _ Stale Eriksen

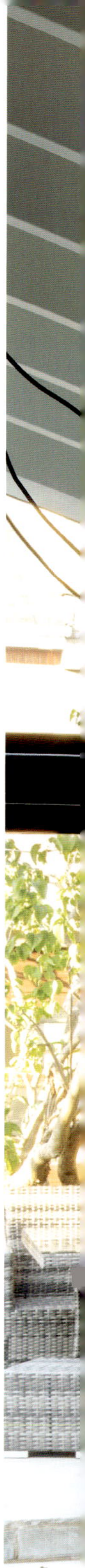

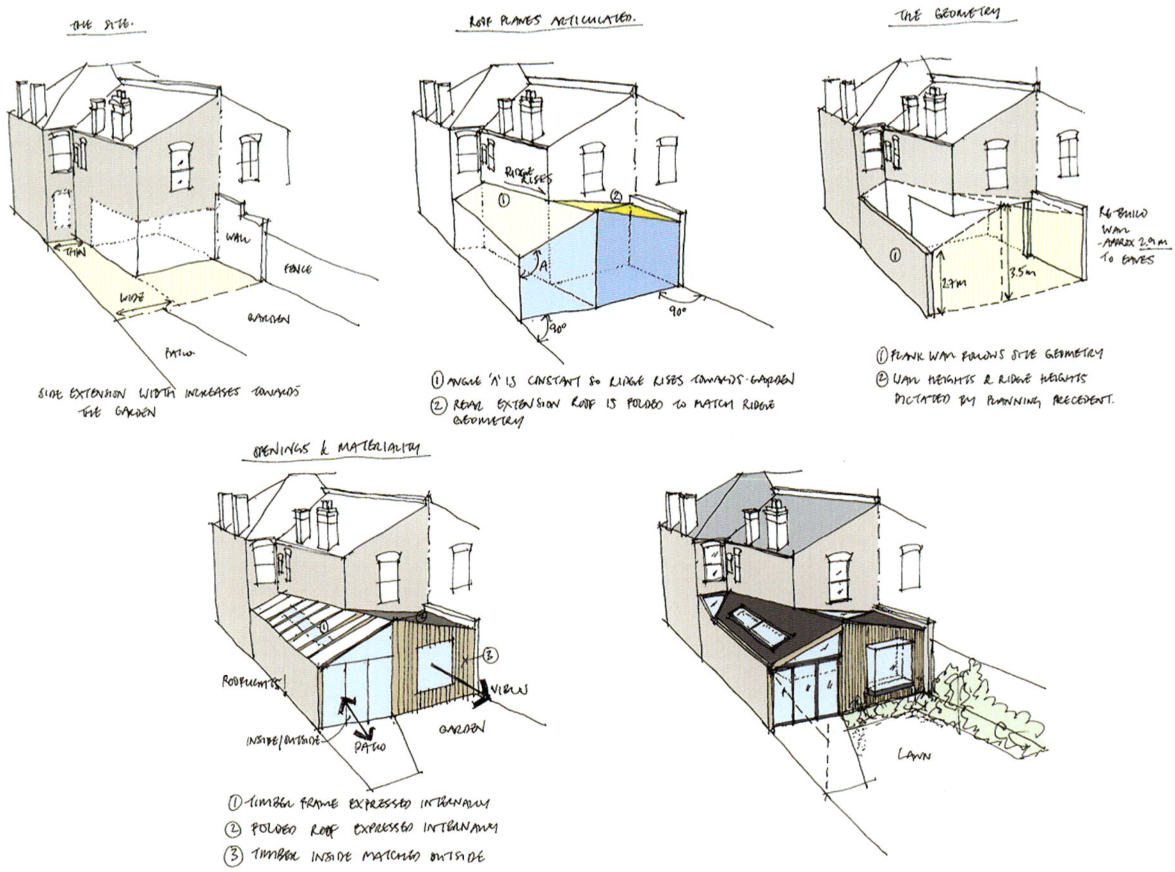

The design of Folded House evolved from a direct response to the geometry of the 'cake-slice' end-of-terrace plot. The brief was to extend as far as was reasonable to create a new open plan kitchen, dining and living space, together with a full program of refurbishment ro the house.

The property was originally end-of-terrace with an angled flank wall that created a 'cake-slice' shaped in-fill to the side of the house.

The site geometry formed the concept for the large rear extension and an angular form evolved. The roof is formed from three 'folds' framed in triangulated steelwork and in-filled with exposed timber framing. The rear elevation is also angled in plan defining the change in material from glass to zinc and function from dining to living, externally from patio to garden.

The angle between the long flank wall and the roof to the side-infill is constant, and therefore the ridge beam rises as the plan widens towards the garden. The roof over the new living space is then folded to resolve the geometry of the angled main ridge beam with the shorter flank wall. An angled custom roof light reinforces the dramatic geometry.

Externally the aluminum cladding is punctuated with an oriel window which frames the garden view from the kitchen. The aluminum cladding is reflected in the dark paneling to the kitchen and utility room wall. Concrete floors, exposed brick, and white painted timber complete the restrained material palette.

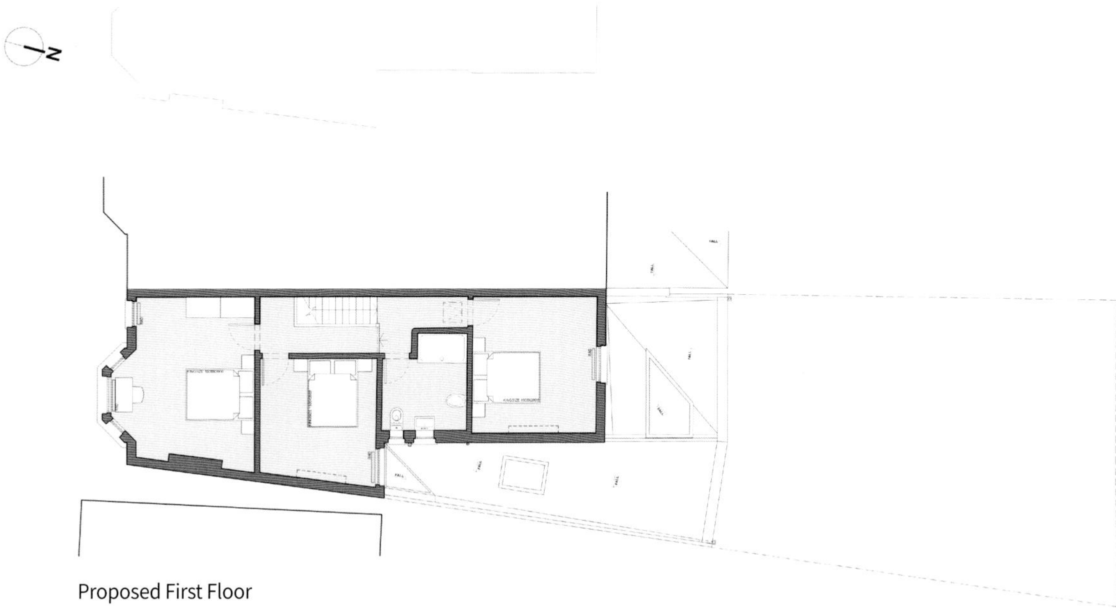

Proposed First Floor

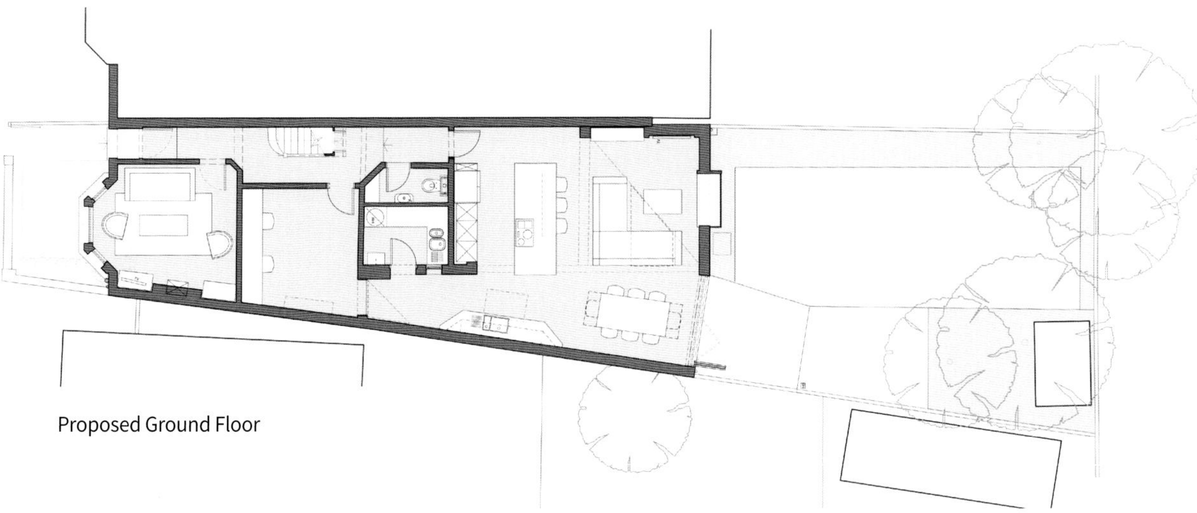

Proposed Ground Floor

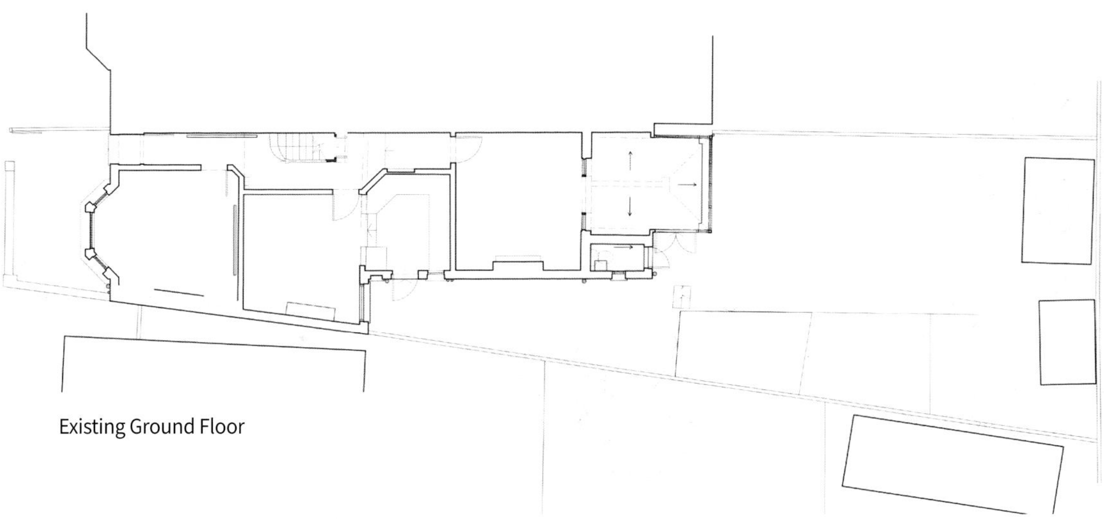

Existing Ground Floor

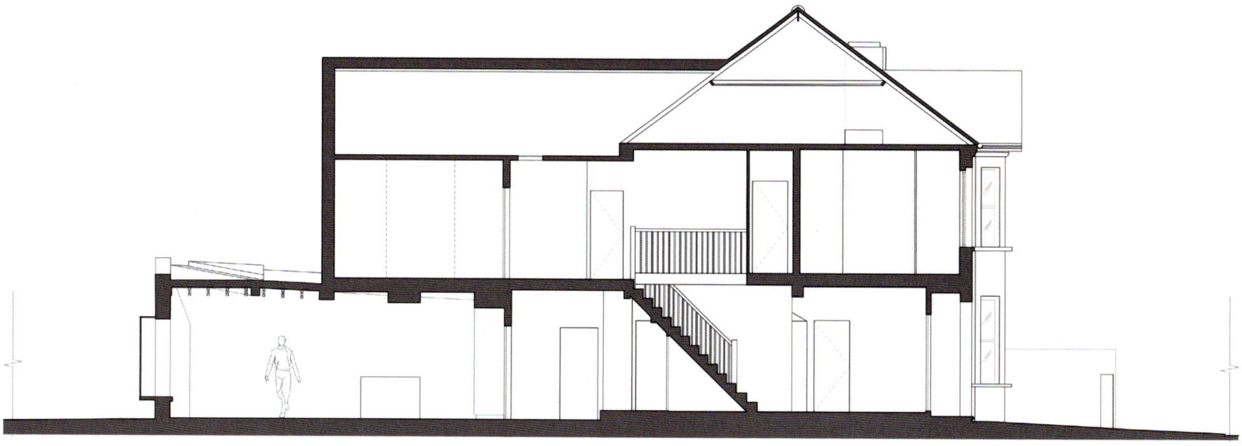

Proposed Section

| **169,41**sq.m. | **Concrete & Steel** |

House for a Photographer

by FORM/Kouichi Kimura Architects

Location _ Shiga,Japan
Site Area _ 420,94m²
Constructed Area_ 169,41m²
Photography _ Yoshihiro Asada, Norihito Yamauchi

303

The site is located along an arterial road that runs through the countryside. In the neighborhood stand private residences and a grove of the village shrine surrounding a passage to the shrine. The building is made of mortar and galvanized steel sheet which reflects dull light, making it look massive. It is laid out to be accommodated within the L-shaped lot and stands facing the shrine on the opposite side. The building is to be used as a studio and gallery of the photographer and his residence as well. The client requested to unite the studio where he produces works with his living space. Thus, I came up with an optimum plan by clarifying how the resident wants to live instead of segmenting the space by function. The dim passage from the entrance approach invites visitors into the innermost space while guiding them with light thrown from the ceiling of the connected gallery. The gallery clips out a landscape with the opening to look like an exhibition. Its contrasted scale and natural light resonate with each other. The space that quietly inspires a feeling of exaltation among ordinary life also plays a role of the photogenic shooting location. The long thin passage serving as an alley to go around the gallery has other functions than a mere pathway; it is also used as a space to post artworks and photos on the wall. It is beautifully themed with sight switching, different ceiling heights, and shade and shadow created by natural light. The hall that is located at the center of the building with different floor levels plays a role of a hub which connects with each space. It also incorporates plumbing equipment, a counter, and a staircase that are necessary for living, and is connected to an approach from the courtyard, thus not only providing functional comfort but also displaying scenic beauty. The building used as a living space also offers a dynamic space with proportion and scale that allow it to serve as a photo studio. The light coming through the high-side light envelops the room, enabling photo-shooting with natural light. Photo shooting equipment, vintage furniture, musical instrument, and artworks are placed here and there, blending in with space. It acts in concert with the photographer's feeling and aesthetic and acts as the base to produce new creativity and activities.

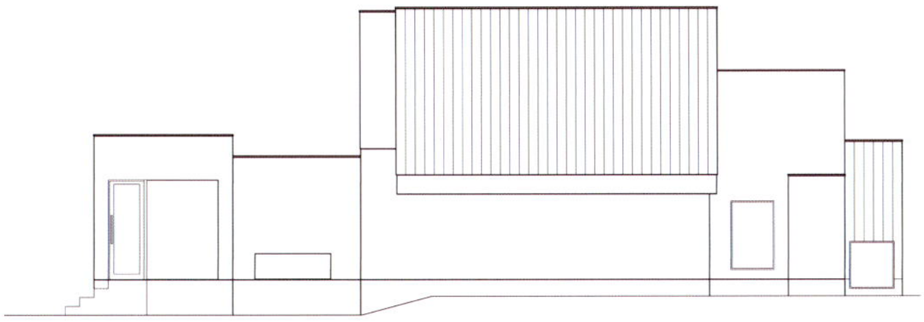

East Elevation

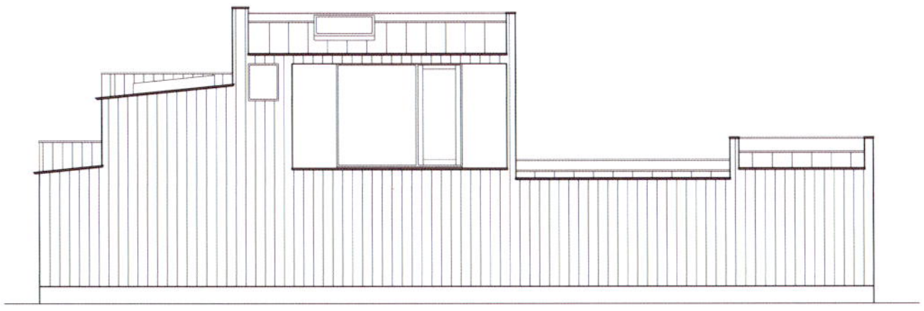

West Elevation

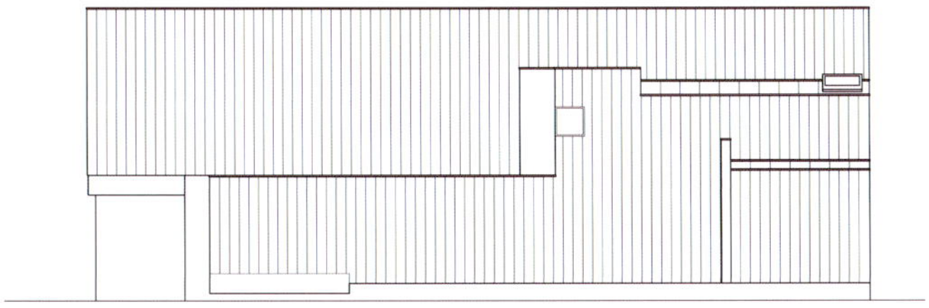

North Elevation

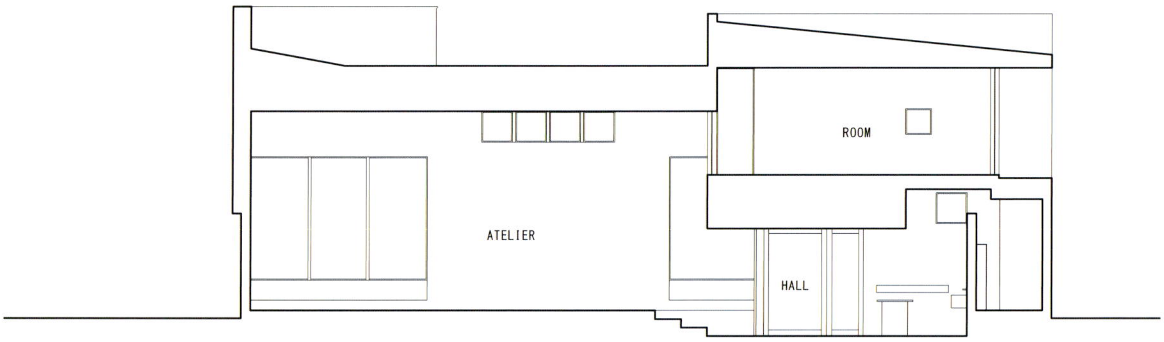

Section A-A'

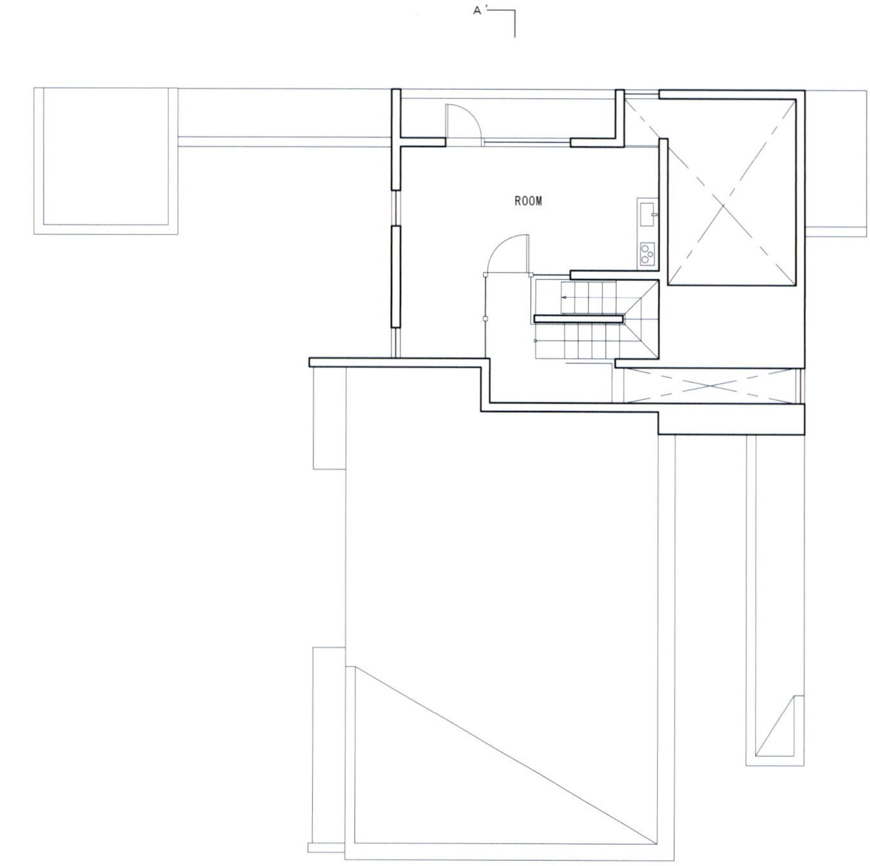

First Floor

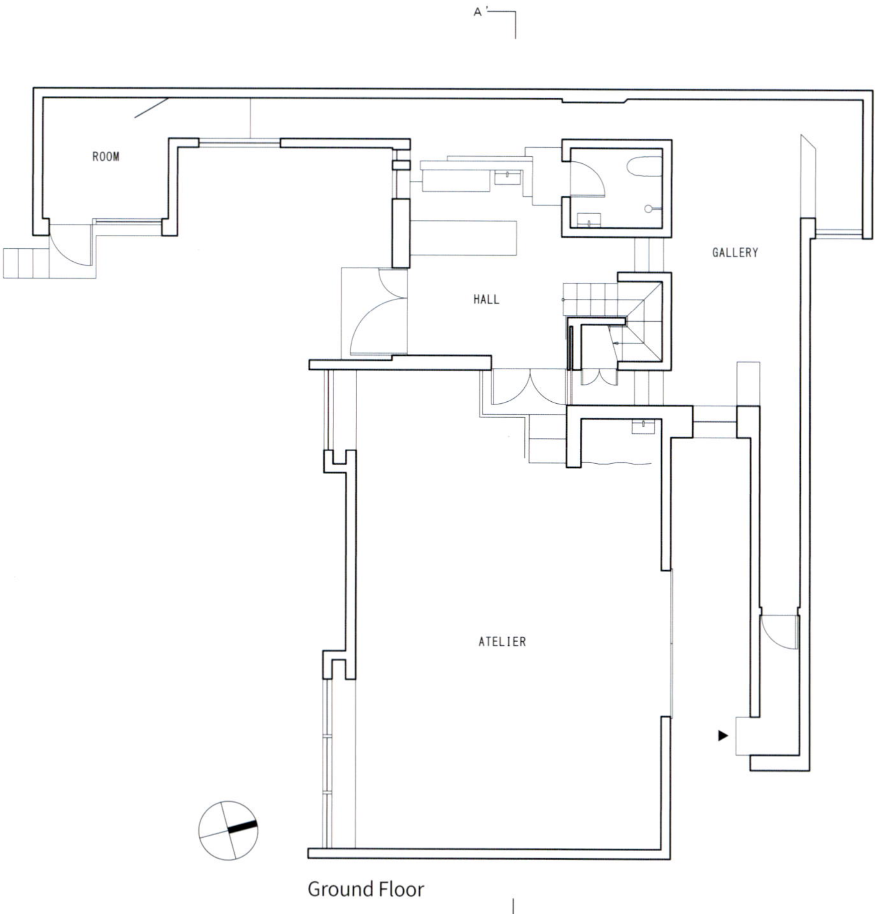

Ground Floor

| **180.41**sq.m. | **Timber** |

SCOPE

by APOLLO Architects & Associates

Location _ Setagaya ward, Tokyo
Site Area_ 144.56m^2
Building Area _ 82.48m^2
Total Floor Area _ 180.41m^2
Structure _ Timber
Scale _ B1F+2F
Structure engineer _ Masaki Structure, Kenta Masaki
Lighting engineer _ Sirius lighting office, Hirohito Totsune
Constructor_ Aoki Koumuten Ltd.,
Photography _ Masao Nishikawa

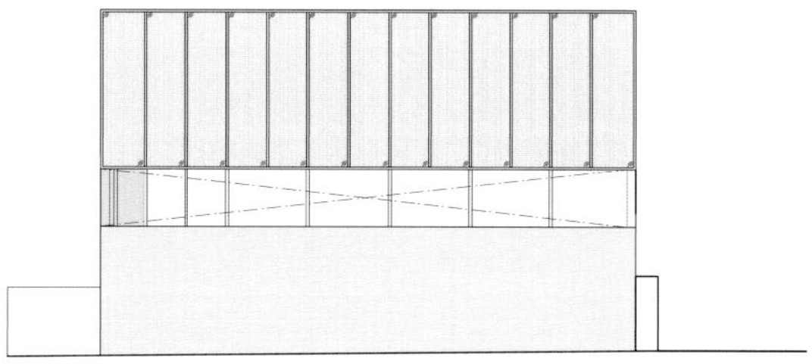
South Elevation

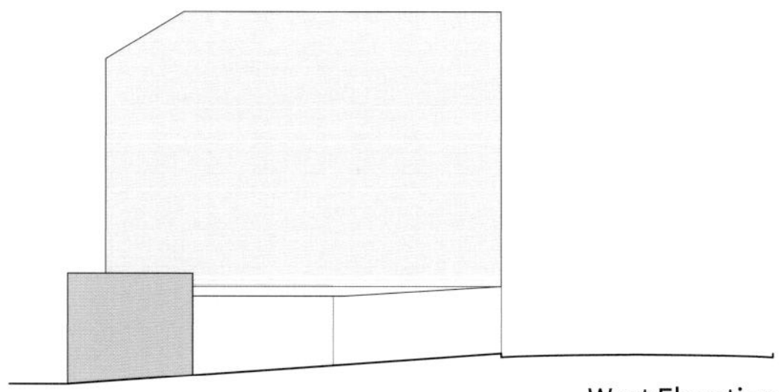
West Elevation

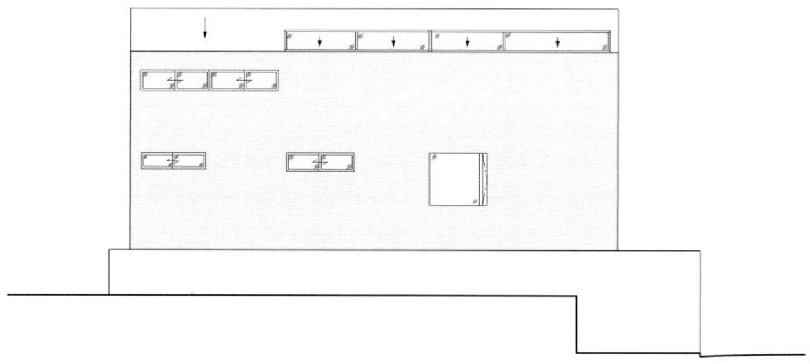
North Elevation

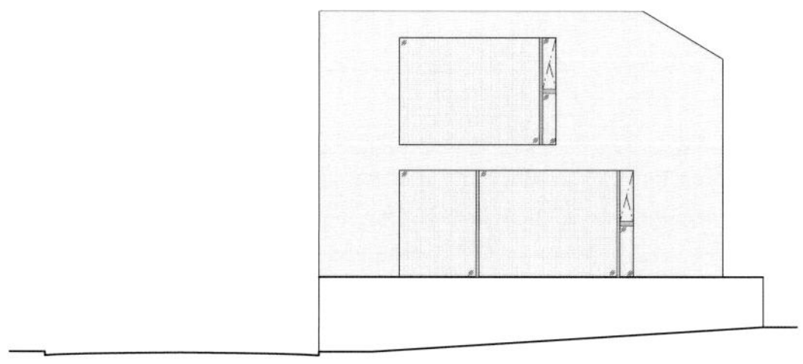
East Elevation

This residence for a two-generation family makes use of the gently sloping site by locating the garage and entryway underground, thereby taking advantage of an allowance in the building code for increased floor area when part of the building is underground. This enables a layout with three stories' worth of floor area, enough for the parents' and children's residences to be fully separate.

Cedar formwork gives the exterior reinforced concrete walls their character, while steel frames surround one-way-mirror windows that maintain both privacy and an open feeling inside. Glass walls separate the garage from the shared underground entryway, making it feel larger than it is. In addition, sunlight pouring in from a skylight lends the entryway a soft ambiance unusual in an underground space.

The parents' residence, located on the first floor, takes advantage of the wide frontage with a deep balcony whose eaves shelter it from rain, making a special place to spend time outdoors. The simple, one-bedroom unit provides plenty of space for the elderly couple. Pine trees on the neighboring property lend a pleasant view to the chic, serene interior.

On the second floor, the younger generation's residence is distinguished by a bright, airy private entryway light up by a skylight on the north side. As in the lower unit, this one includes a balcony, but continuous one-way-mirror windows surrounding the exterior create a penthouse-like feeling of openness. The overall impression contrasts with that of the under-eaves balcony below. In the evening, light from both units beautifully illuminates the building and enlivens the quiet streetscape.

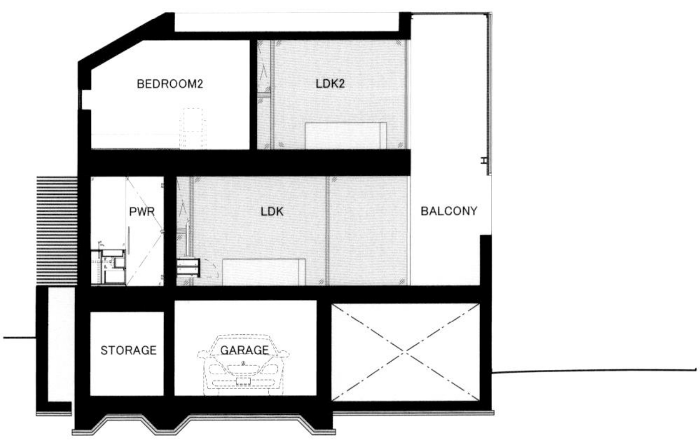
BEDROOM2
LDK2
PWR
LDK
BALCONY
STORAGE
GARAGE

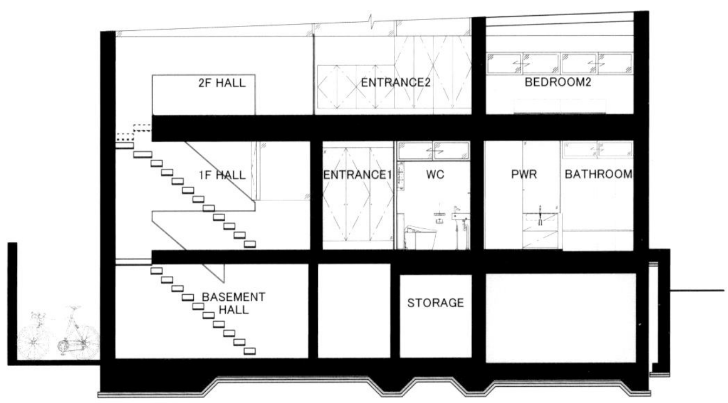
2F HALL
ENTRANCE2
BEDROOM2
1F HALL
ENTRANCE1
WC
PWR
BATHROOM
BASEMENT HALL
STORAGE

2F HALL
BATHROOM
MASTER BEDROOM
1F HALL
BEDROOM1
BALCONY
BASEMENT HALL
GARAGE

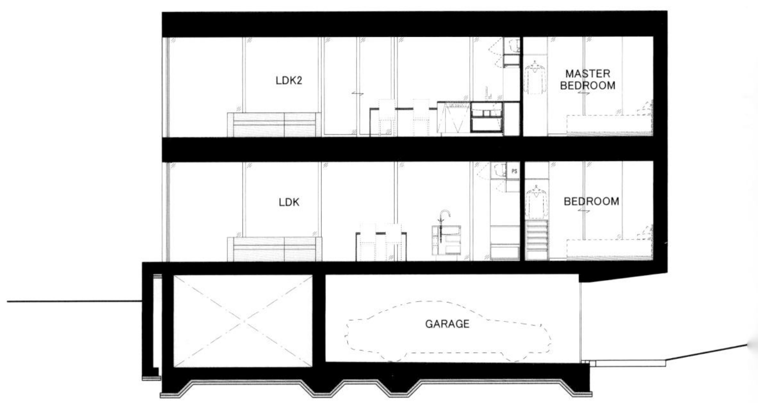
LDK2
MASTER BEDROOM
LDK
BEDROOM
GARAGE

Sections

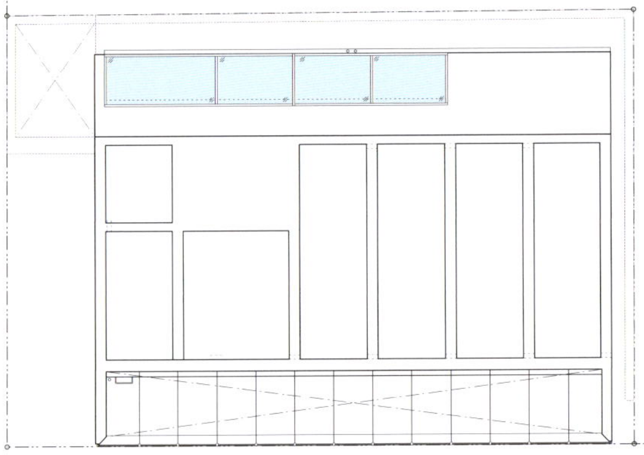
Roof Floor

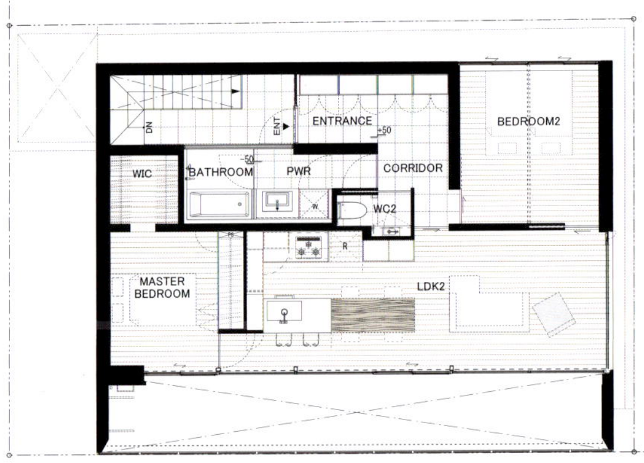

First Floor

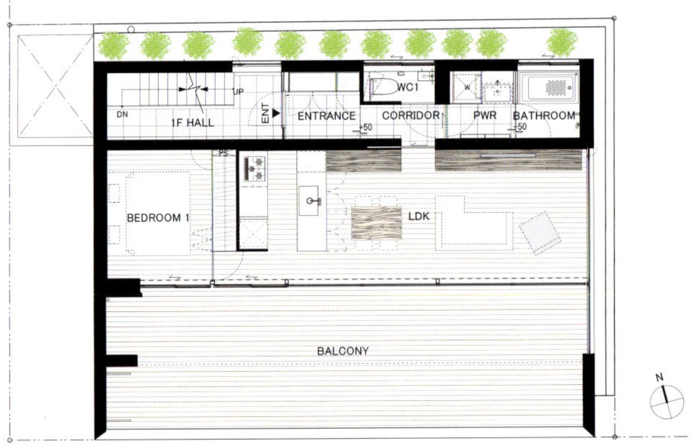

Ground Floor

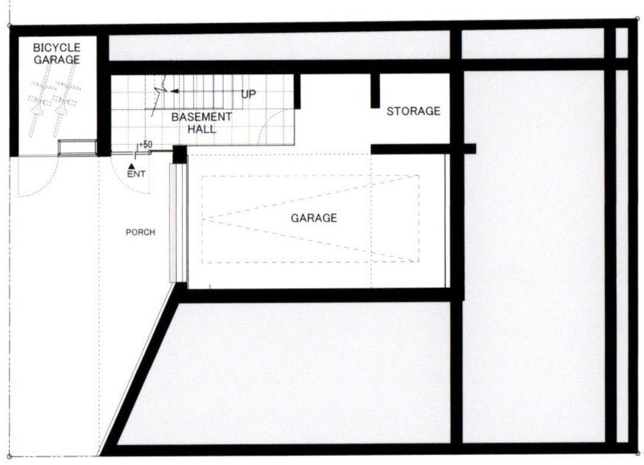

Basement Floor

Participation Designers_

0123456789abcdefghijklmnopqrstuvwxyz

Abarca + Palma Arquitectos p.080

Francisco Abarca and Camilo Palma are Chilean architects who manage the architecture office "Abarca + Palma". This Office exists since 2009, the year in which they received the first assignment to design and build Casa Lo Cañas (2009) in the commune of La Florida in the city of Santiago de Chile. From this project the architects install a way of thinking, designing and constructing the work from a perspective that has led them to develop a continuous work in time where the most relevant factors have been material expression, economy and a repertoire of spatial and formal arguments that are linked to the observation of the built landscape of Chile. Mainly the work has been focused on the development of housing as an architectural typology and construction in wood as a construction typology both linked to the economic possibilities that each project presents, resulting in a catalog of low-cost works. Among the most relevant factors in spatial terms is the constant search for the intermediate space in various forms and in various relationships with other spaces whether interior or exterior, generating a constant link between interior and exterior by means of patios, corridors, large eaves or simply covered open spaces that relate the inhabitant to their environment.

Francisco Abarca (1981) / *Right side*
Architect Andrés Bello University, 2006. Diploma in Sustainable Architecture Universidad de Chile 2007.

Camilo Palma (1979) / *Left side*
Arquitecto Universidad Andrés Bello, 2006. Profesor de Taller de Arquitectura Universidad Andrés Bello Campus Creativo.

Akasaka Shinichiro Atelier p.024, 192 / p.046

Shin-ichiro Akasaka

1970 Born in Sapporo,Japan
1993 Graduated from Hokkai Gakuen University, Fuculty of Engineering, Architectural Department
1993~ Yoshimi Nakai Architect & Associates
2000 Established Akasaka Shinichiro Atelier

2018 JIA Selected Works
2012 INAX Design Contest Gold medal
2011 AIJ Hokkaido architecture prize for encouragement
2011 Wood Application Competition, Special Prize
2010 Hokkai Koki Co., Design Award, Grand prix
2010 JIA Hokkaido Branch Residencial Building Award, Grand prix
2009 Selecded Architectural Designs AIJ
2006 Wooden Architectural Space Design Competition, Residential Design Award
2006 JIA Selected Architectural Designs, Awarded
2005 JIA Selected Architectural Designs, Awarded
2005 JIA Hokkaido Branch Residencial Part, New Hope Prize
2003 JCD Design Award, First Prize
2002 JCD Design Award, Encouraging Prize
Lecturer at Hokkai Gakuen University
Lecturer at Hokkaido College of Art & Design
Lecturer at CAI Contemporary Art Institute
Member of Japan Institute of Architects
Member of Architectural Institute of Japan

Aki Hamada Architects p.186

Aki Hamada

1984 Born in Toyama prefecture, Japan
2010 Tokyo Metropolitan University, BA Architecture and Urban Design
2012 University of Tokyo, MA Architecture
2012 Established studio_01 with Alex Knezo
2014 Founded Aki Hamada Architects
2014 Joined teamLab Architects as partner
2014~2016 Lecturer at Nihon University

Alejandro Soffia p.062

Alejandro Soffia has been developing long term R + D since 2002. He focuses on material optimization and prefabrication, as a way to improve construction impact on the environment and give access to good design, to a wider range of the population. Projects have been done on different scales and destinations, but they have been focused on materials, techniques, and details. This sort of interest has been a perfect complement in designing building systems, even though these systems not always are prefabricated, but give high flexibility to their users. Alejandro Soffia has taught on Design, Architecture and Graduate Schools in Chile. Many of my projects have been published on international Architecture Magazines like DOMUS (Italy), AMC Le Moniteur (France), ARQ/CA/SPAM/AS/180 (Chile), 2G, Arquitectura Viva (Spain), Arhitectura (Romania). Also been part of some books like "Atlas of 21st Century Architecture" (Phaidon), "Wood Architecture Now II" (Taschen), "21st Century houses" (Abbeville), "1000x Architecture of the Americas" (Braun).

Alphaville Architects p.258 / p.278

Kentaro Takeguchi

1994 graduated from School of Architecture, Faculty of Engineering, Kyoto University
1995~96 studied in Architectural Association School of Architecture, U.K., in Diploma Unit 5, Foreign Office Architects
1998 graduated from master course, School of architecture, Kyoto University
1998 established ALPHAVILLE Architects with Asako Yamamoto
2012 Adjunct professor at Osaka Sangyo University(-2017), Lecturer at Kobe University
2018~ Visiting professor at Kobe University, Lecturer at Ritsumeikan University

Asako Yamamoto

1994 graduated from School of Architecture, Faculty of Engineering, Kyoto University
1995~96 studied in l'Ecole d'Architecture de Paris, La Villette, France
1997 graduated from master course, school of architecture, Kyoto University
1997 worked at Riken Yamamoto & Field Shop
1998 established ALPHAVILLE Architects with Kentaro Takeguchi
2010~ Lecturer at Kyoto University, Osaka Institute of technology

ALPHAVILLE Architects Co.,Ltd, founded in 1998 by Kentaro Takeguchi and Asako Yamamoto, is a 10-person firm based in Kyoto JAPAN, engaged in architectural projects of any scale. We always try to find out the 3-demensional structure of the space, which is mainly consist of the minimum opening and the structure itself. We are asking the effectiveness of the architectural space for human activity and make best use of geometry with the sustainability, calculating minimum material manipulation for diverse geometrical spatial results. Recently we are trying to adapt our experiment to lager scale like catholic Suzuka church project.

ALTS Design Office p.084, 178 / p.012, 118, 204

Sumiou Mizumoto

Born on April 13, 1977 in Japan
The registered architect (second class)
March,1997 Graduated from a vocational school for architectures (Japan)
September,1998-October, 2009 Architectural design office "Kankyo design Co.ltd" (Japan)
November,2009- January, 2012 architectural design office (Japan)
February, 2012 Established ALTS DESIGN OFICE
August, 2014 Corporateizationed to ALTS DESIGN OFICE Co.ltd

2017 LIXIL members contest 2017, the fighting spirit award of new construction section, "Otsu house"
2017 Best of house 2017, Design award, "Ishibe house"
2016 Best on Archilovers 2015, Winning a place, "Ishibe house"
2016 4th house construction prize, Award of general vote, "Higashihayashiguchi house -Enjoying storage- "
2016 Best of house 2016, Design award, "Kurihigashi house"
2015 Interior coordination contest of living, Section award of new construction section, "Hazukashi house"
2015 3rd house construction prize, Townscape award of judges selection section, "Agui house"
2014 LIXIL members contest 2014, Regional excellence award of new construction section, "Suehiro house"

APOLLO Architects & Associates p.230 / p.122 / p.096, 164, 214, 222, 256, 314

Satoshi Kurosaki (President of APOLLO Architects & Associates, Ltd)1970 Born in Sapporo,Japan

Kurosaki established APOLLO in 2000 that is the architecture studio from the professional career experience of the product development in the large residential company. APOLLO's offices are in Tokyo, Okinawa, Korea and China, and the primary works are residential, villa, resort, clinic and commercial facility. Kurosaki is one of the most practicing architects who designed over 150 buildings for over 16 years in Japan. He also published a full-coloured 600 pages APOLLO's Monograph (Germany) in 2016.

/ First-class architect General Design Department of Architecture, The Japan Institute of Architects.
/ Asia-Pacific Economic Cooperation ARCHITECT.
/ Graduated from the Faculty of Architecture, Meiji University.
2008~2010 Nihon University Part-time lecturer
2014~ Keio University Part-time lecturer

Awards Received

GOOD DESIGN AWARD (JAPAN) / Tokyo Association of Architectural Firms / The Japan Institute of Architects Excellent Architectural Award / Modern Living Award JAPAN / HISE Design Award(Slovenia) / Wallpaper Design Award(England) / BUILD Architecture Awards (England) / Architecture Awards for 2016 (England) / GERMAN DESIGN AWARD (Germany) / ICONIC Award (Germany) / Design for Asia Award (Hong Kong) / INTERNATIONAL SPACE DESIGN AWARD (China) / DNA Paris Design Award (France)

archimania p.100

Each archimaniac contributes a unique range of knowledge, talents and skill sets to form the creative and collaborative environment at archimania. The common threads, linking and driving these diverse design professionals is a passion for architecture and a commitment to improving the built environment.

Atelier Tsuyoshi Tane Architects p.270

Tsuyoshi Tane

Tsuyoshi Tane is a Japanese architect based in Paris. He founded Atelier Tsuyoshi Tane Architects in 2017 after working in London, Denmark and Japan, and co-founding DGT in 2006. Through a series of award-winning projects and buildings such as Estonian National Museum, Kofun Stadium for the Japanese Olympics 2020, Todoroki House in Valley, Furoshiki Paris and many others; Tsuyoshi has become recognized in a new generation of emerging architects. He believes architecture belongs to a memory of place and can carry the past into the future. It is his concept - "Archaeology of the Future". Tsuyoshi has received numerous awards and honours, including the French Ministry of Culture Architecture Prize, French Architects Overseas Grand Prix 2017, Estonian Cultural Endowment Grand Prix, nomination for the Mies van der Rohe Award 2017 and the 67th Japanese Ministry of Education for Fine Arts New Face Award. He frequently gives public lectures and has been teaching at the Columbia University GSAPP and ESVMD.

bergmeisterwolf architekten p.130

bergmeisterwolf is a design and architecture studio based in brixen and rosenheim that for 13 years has been offering answers to contextual and spatial requests through an intense dialogue with its clients and the academic research. the studio was established by gerd bergmeister and michaela wolf, who work in close collaboration with architects, artists and a network of professionals from different fields. bergmeisterwolf works include architecture and interior design in hospitality, residential, cultural and commercial buildings, both private and public. the singularity of each intervention results from the measured pursue for a new balance in the territory: bergmeisterwolf re-discovers antique construction techniques while introducing and experimenting with new ones; it recognizes and enhances the continuity of existing buildings and, at the same time, it heightens them with new discontinuity elements, it recovers the shapes of the place and is contaminated by new contemporary languages, it builds underneath the existing morphology and transforms the landscape. bergmeisterwolf works with the past, the present and the future as a way to understand and reveal the identity of the place.

BUREAU p.134 / p.020

Daniel Zamarbide (Architect FAS – IAUG - SIA | BUREAU / Co-director of ALICE laboratory at EPFL)

Daniel Zamarbide obtained his master degree at the Institut d'Architecture de l'Université de Genève (IAUG) in 1999. During his studies, he followed the workshops of Christian Marclay, Philippe Parreno and Catherine Queloz at the Ecole Supérieure de Beaux Arts in Geneva. In the year 2000 he becomes a founding member of group8, an architectural practice that has acquired an important national and international recognition. Daniel Zamarbide has developed through the years a particular interest in the protean aspects of his discipline and nourishes his work and research through other domains like philosophy, applied and visual arts as well as cinema. As a guest lecturer and jury, he is often invited to a diversity of international architectural schools and cultural institutions to present and discuss his work and researches. Since 2003 his interest in education has led him to be a professor at the HEAD (Haute École d'Art et Design) in Geneva and a guest professor and co-director of ALICE laboratory's year one at the EPFL (École Polytechnique Fédérale de Lausanne). In 2012 Daniel left the group8 to pursue a more personal path, closer to his more intimate interest. His project (now signed under the generic name of BUREAU), explore the possibilities of architectural realm. Architecture is thought and practiced in a variety of formats (scenography, installations, objects, designs, permanent and temporary architecture). Critical writing is also part of Daniel's activities which he develops in different institutional publications as well as his own journal, an editorial project eponymous to his practice that has begun in 2017.

CDM Casas de México p.154

Our experience of more than 20 years designing, building and setting the homes of diverse families in Latin America and the United States, has consolidated us as a residential architecture firm with an international presence. Our main goal is to ensure that you inhabit a space with which you can identify, that projects your lifestyle and that meets the requirements of an optimal quality of life. We seek to create "habitable experiences" through spaces that evoke the spirit of the person who inhabits it, making the home what all homes should be: the extension of oneself.

Javier Dueñas / *Left side*
Jaime De Obeso / *Right side*

Christoph Hesse Architects p.056, 236, 302

Christoph Hesse, born 1977, grew up in the highland area of Sauerland, Germany. He received a Diploma in Architecture from ETH Zurich and a Master of Architecture in Urban Design with Distinction from Harvard University. He is the founder of Christoph Hesse Architekten, an award winning design fi rm based in Korbach and Berlin, Germany. Before, he worked in renowned offices in Frankfurt, Berlin and London. In 2018, Christoph Hesse received the European 40 under 40 Architecture and Design Award and was nominated for the DAM Preis 2019. His project „Villa F" was shortlisted for the DAM Preis 2018, honouring the 20 best projects of the year by the German Museum of Architecture. Additionally in 2017, he awarded the American Architecture Prize, the International Architecture Award, the Architizer A+ Award and was shortlisted for the Bauwelt Award. In 2016, he received the Iconic Award, the German and Austrian Façade Award and the Boston Society of Architects Design Citation Award. Projects by Christoph Hesse Architekten have been featured and published in the German Architecture Annual, the Metropolitan Arts Press, Bauwelt and Stadtbauwelt, New International Architecture, Architecture Guide of Germany, MARK Magazine, Frankfurter Allgemeine Zeitung, The Best of Architizer by Phaidon, Stylepark Magazin and ZDF Aspekte. His projects have been exhibited at Contemporary Space Athens, Experimenta Urbana Kassel, German Museum of Architecture and Museum für Angewandte Kunst in Frankfurt, Architecture Biennale in Venice, World Architecture Festival, Kraftwerk Mitte, Architektur Galerie Berlin and New Museum of Contemporary Art in New York. Christoph Hesse lectures regularly at conferences and universities worldwide, e.g. Columbia University New York, IIT Chicago, Tongji University Shanghai, Hong Kong University, National University of Singapore or Strelka Institute Moscow. He has taught architecture and urban design at University of Kassel, TU Darmstadt, University of Architecture Saigon, Harvard University, and Cairo University.

FORM / Kouichi Kimura Architects p.126, 142, 154, 302

Kouichi Kimura

1960 Born in Kusatsu City, Shiga prefecture, Japan
1982 Graduated from Kyoto Art College
1994 Established FORM / Kouichi Kimura Architects

My works, mainly residences, are characterized by a poetic and functional Minimalism in which a sense of openness and privacy is achieved through Careful slicing and layering of wall elements,light and shadows,volumes and spaces.

Daniel Moreno Flores p.122 / p.310

Jose Maria Saez Vaquero

Born in Avila, Spain in 1963. Architect graduated from the Polytechnic University of Madrid. Specialist in Environment and Bioclimatic Architecture and Conditioning and Installations for Architectural Rehabilitation. Winner of the award for the Best Young Work, on the VI Ibero American Architecture and Urbanism Biennale of Lisbon in 2008, for Pentimento House, that also received the National Award for Architectural Design in the XV Pan-American Architecture Biennale of Quito in 2006. Designer and constructor, standing out the rehabilitation in Quito of his own house Saez Moreira House, winner of the Heritage Award for Architectural Restoration in the Historic Center of Quito in 2005, and the San Juan House: National Honorable Mention for Architectural Design in the XVI Pan-American Architecture Biennale of Quito in 2008 and Selected Work on the VII Ibero American Architecture and Urbanism Biennale of Medellin in 2010. National Award for Architectural Design in the XVIII Pan-American Architecture Biennale of Quito in 2012 for Algarrobos House. Founder teacher at the School of Architecture, Design and Arts of the Catholic University of Quito. Director for Architecture from 2001 to 2002. Invited teacher in universities of Ecuador, Argentina, Peru, Spain and United States.

Daniel Moreno Flores

Born in Marsella, France in 1984. Architect graduated from the Pontificia Universidad Católica del Ecuador of Quito. He is Studing a master in Argentina in Advanced Architectural Design. The Graduation Project was Winner of the Studens award Glod Medall in the XVI Pan-American Architecture Biennale of Quito in 2008. This project also participated in the Archiprix Montevideo 2009, world´s best graduation projects. Mikes Study obtained the National Honorable Mention for Restoration and Recycling in the XVIII Pan-American Architecture Biennale of Quito in 2012 and the Algarrobos House received the National Award for Architectural Design in the XVIII Pan-American Architecture Biennale of Quito in 2012.

Extrastudio p.288

Extrastudio is a Lisbon based architecture practice established in 2003. The office develops small and large projects with programs ranging from housing to retail and public space. In recent years we have received public commissions for the design of waterfront and urban parks, and have also worked with private clients on the design of mixed used buildings and houses.

FujiwaraMuro Architects p.162, 290, 298 / p.202, 226 / p.134, 196

Shintarou Fujiwara / *Left side*

1974 Born in Osaka, Japan / 1997 Graduated from KINDAI University, Faculty of Engineering, Department of Architecture / 1999 Degree of M.Eng. At KINDAI University, Graduate School of Science and Enginnering Research / 2002 start FujiwaraMuro Architects

Yoshio Muro / *Right side*

1974 Born in Nagoya, Japan / 1999 Graduated from KINDAI University, Faculty of Engineering, Department of Architecture / 2002 start FujiwaraMuro Architects

GENETO p.238

GENETO is a creative unit working on variety of projects from Urban design, Architecture, Product design to Art installation. GENETO blends these to seek the new imagination of creative value. GENETO specializes in designing, including all kind of architecture design as well as interior and art installation. To work in cooperation with pivoto, our group company which specializes in furniture production, we are able to achieve a consistent quality and service throughout project. Working on variety of projects without setting any boundaries, we aim to generate new value that yet to be found.

gualano+gualano arquitectos p.220 / p.214

Marcelo Gualano / Montevideo, 1967 / *Left side*
Architect / Facultad de Arquitectura - UdelaR
Professor Agregado de Proyecto grado 4 / Farq / Udelar
Co-director Estudio Glot Diseño / 1999-2009
Co-director revista de arquitectura dEspacio : 2004-2005
Member of Editorial board Revista R / Facultad de Arquitectura 2012 - 2017

Martín Gualano / Montevideo, 1969 / *Right side*
Architect / Facultad de Arquitectura - UdelaR
Professor Adjunto de proyecto grado 3 / Farq / Udelar
Construction Manager / DNA:MTOP : 1995-2005

Henning Stummel Architects p.176

Prior to setting up the practice in 2000, Henning worked with Sir David Chipperfield (1993 to 2000) and previously with Lord Foster (1992-93). He studied at the prestigious TH Darmstadt, ETH in Zurich. He was born in 1966 and has lived in Germany, Switzerland and the UK. Henning has lectured and taught at various places, including directing a design unit at Cambridge University. His work has taken him around Europe, giving lectures and presentations. He has a general interest in all aspects of design – ranging from consumer products to the pressing issues raised by rapidly expanding conurbations. Fascinated by simple, functional design, there is a great appreciation of the things that transcend time – the reduction, order, serenity and tranquillity. Henning (and his wife Alice), love traveling with their sailboat. This tranquil pursuit combines the joy of discovery, the love of good food and the appreciation of the minimal.

Ikeda Yukie Ono Toshiharu Architects p.268, 276 / p.05

Yukie Ikeda / *Left side*

1973 Born in Miyagi Prefecture
1998 Bachelor of Architecture, Technology Department of University of Tokyo
1999-00 Exchange Student at Helsinki University of technology,Finland
2001 Master of Architecture, Technology Department of University of Tokyo
2001-03 Shigeru Ban Architects
2001-03 Temporary Lecturer, Intermedia Art Department of The National Art University
2003-05 studio Archi Farm
2006 Yukie Ikeda Architects, First-Class Registered Architect Office in Japan
2013 Temporary Lecturer, depertment of architecture, Tokyo University of science
2014- Temporary Lecturer, depertment of architecture, Sagami Women's University
2017- Temporary Lecturer, depertment of architecture, Kyoto University of Art and Design
2017 ikedayukie onotoshiharu architects, First-Class Registered Architect Office in Japan
2018- Temporary Lecturer, depertment of architecture, Kokushikan University
2019 Temporary Lecturer, depertment of architecture, Tokyo University of science

Toshiharu Ono / *Right side*

1978 Born in Tokyo
2001 Bachelor of Architecture, Waseda University
2004 Master of Architecture, Waseda University
2004 Schemata Architects
2008 Yukie Ikeda Architects
2013 Yukie Ikeda Architects, Partner
2014- Temporary Lecturer, depertment of architecture, Sagami Women's University
2017 ikedayukie onotoshiharu architects, First-Class Registered Architect Office in Japan

Kenta Eto Architects p.046, 096

kenta Eto

1979 Born in Oita Japan
2001 Kyoto International Architecture College
2011 Kenta Eto Atelier establishment
2012 Good Design Award 2012
2012 AICA Award 2012
2013 Good Design Award 2014

Life style Koubou p.046

We are always impressed and fulfilled by the four beautiful seasons of Mt Adatara. Having grown up here, surrounded by Mt Adatara, we have learned the wisdom of living. This important wisdom, which has been passed through the generations, has both eternal and universal value. We believe that we design with this wisdom, not only here but worldwide. To do so, we firstly examine both the climate and nature of a specific area with care. After that, we propose ideas that locals will love and support. As we look towards the future, we keep challenging ourselves. How many ideas we can realise that will be breath-taking and exciting? How many projects can we design worldwide that will benefit our hometown of Adarata? We would like to show 'the lifestyle' of each locale through our unique ideas and designs, which emerge when we are thinking and creating while taking in the view of Mt Adarata from our office

Marilia Pellegrini Arquitetura p.092

Marilia Pellegrini graduated in Architecture and Urbanism at the Federal University of Minas Gerais (UFMG) in 2001. Her professional experience began with the architect Carlos Alexandre Dumond, Carico. In 2004, Marilia moved to the United States to work at Rene Gonzalez Architect's office, renowned for its residential projects and contemporary lines buildings. Back in Brazil in 2006, Marilia joined the team at Studio Arthur Casas, where she remained until 2015. With a contemporary look, as a result of her experience and constant research, Marilia develops, along with her team, architectural projects and commercial, residential interiors and also explores furniture design.

Meguro Architecture Laboratory p.074, 244

Mitsuru Hirai / *Left side*

(b. 1974, Hokkaido, Japan) Master course at the Department of Architecture, Kogakuin University. Withdrew from the doctoral program at the Graduate School of Architecture, Kogakuin University, upon earning all required credits in 2009. Established Drawing Notes Architectural Design Office in 2009. Changed the name to Meguro Architecture Laboratory in 2015. He is now a part-time lecturer at Jissen Women's University.

Sayu Yamaguchi / *Right side*

(b. 1985, Tokyo, Japan) Doctoral Course, Graduate School, Japan Women's University. Established Drawing Notes Architectural Design Office in 2009. Changed the name to Meguro Architecture Laboratory Co., Ltd. In 2015. Now she is a part-time lecturer at Tokai University and Japan Women's University.

Casper Mork-Ulnes

Lexie Mork-Ulnes

Mork-Ulnes Architects p.192

With offices in San Francisco and Oslo, Mork Ulnes Architects approaches projects with both Scandinavian practicality and California's 'can-do' spirit of innovation. Rigorous and concept-driven, the practice is based on built work characterized by both playfulness and restraint, and informed by economies of means and materials. Since its founding by Casper Mork-Ulnes in 2005, Mork Ulnes Architects have worked on projects from master plans, to mixed-use buildings, to ground up residences, to 100 square foot cabins, and have built on 3 continents. Our current projects include: a master plan for a new community in the California Sierras; a mixed-use, waterfront community in the Bay Area; a cross-laminated timber home in Oregon; a surf cabin in Brazil; and a 110,000sf mixed-use building in Norway. Mork Ulnes Architects has been the recipient of numerous national and international honors, including Architectural Record's 2015 Worldwide Design Vanguard award. The firms' work has also been widely featured in international publications such as The New York Times, Wallpaper, Mark, and Dwell Magazine.

Naf Architect & Design p.180

naf architect & design was established in 2000 by Akio Nakasa and Tetsuya Nakazono, who both studied architecture at Hiroshima University. Prior to its establishment, Nakasa earned his master's degree at Waseda University in Tokyo and worked at Riken Yamamoto. Since 2001, Nakasa has been Tokyo Office representative mainly working with clients in eastern Japan, while Nakazono continued in Hiroshima to work on projects in western Japan. They have designed a wide range of projects from private home, clinics, nursery schools and apartment buildings, all of which are concept oriented reflecting the philosophy of the clients. Their goal in design is how best architectural expression and solution are materialized in the project. Among many awards received, "small village", nursery school, received GOOD DESIGN, JIA award, Kids Design and exhibited during Milano triennale, and "A House Made of Two", private home, and, "Machida Shizen Kindergarten" received JIA award.

OFIS arhitekti p.036

OFIS arhitekti is an architectural practice established in 1996 by Rok Oman (1970) and Špela Videčnik (1971), both graduates from the Ljubljana School of Architecture (graduated October 1998) and London's Architectural Association (MA January 2000). By 1998 they had already won several prominent competitions, such as Football Stadium Maribor and City Museum Ljubljana. Many of their projects have been nominated for a Mies van der Rohe award, a Silver IOC/IAKS medal for the football stadium in 2009, the European Grand Prix for Innovation Award in 2006, an honorable mention at the Miami Bienal in 2005 for their Villa "Under" Extension, a high commendation for the City Museum renovation and extension by the UK Architectural Review's AR+D awards, and the prestigious "Young Architect of the Year" award in London, UK, in 2000, to just mention a few of their achievements.

People's Architecture Office p.012

In 2010, He Zhe, James Shen and Zang Feng co-founded Beijing-based People's Architecture Office, a multi-disciplinary studio focused on social impact through design. PAO is the first architecture practice in Asia certified as a B-Corporation and serves as a model for social entrepreneurship. The office of PAO is located in a traditional courtyard house in Beijing's historic core, a setting characterized by urban informality, and one that inspires the studio's work. PAO engages in urban issues through designs that straddle architecture and product design. Fast Company listed PAO as one of the world's ten most innovative architecture companies in 2018. The studio's award-winning works have been exhibited at the Venice Architecture Biennale, Harvard Graduate School of Design and the London Design Museum.

Proctor & Shaw p.142 / p.292

John Proctor and Mike Shaw have collaborated together since 2004 and share a passion for the quality of design and professionalism required to deliver high quality architecture. Our collective experience spans small and large projects across a range of sectors including crafted individual homes, high quality commercial work and award-winning public buildings. We have previously worked with some of the UK's most recognized cultural institutions such as the V&A museum, The Wallace Collection and The Design Museum.

RVArchitecture p.246

Our design process starts with the existing environment. We are inclined to the local, the small scale, but we do not merely copy it. We are averse from architecture that is associative. We aim to design in the pure grammatical language of architecture. At the end of the day it is our ambition to make contemporary architecture that intensifies the character of the existing environment. Ruud Visser Architects is located in a small village in the middle of Holland. From this typical Dutch landscape we study the specifics of locations. Sometimes it concerns a typological study of houses built in a specific village. Other times it concerns the typological study of small bridges that connect houses to the street in a certain region. But it can also consist of a study to the possibilities of reuse of an existing building. These studies form the fundament underneath the architecture of our office.

Sanden+Hodnekvam Architects

p.060 / p.282

Sanden+Hodnekvam Architects is an architectural studio based in Oslo/Nesodden, Norway. The studio was established in 2014 by John Sanden and Ingvild Hodnekvam and is engaged in a varied range of projects all over the country.

Shogo ARATANI Architect & Associates

p.306

SHOGO ARATANI

1971 Born in Osaka
1996 Graduate from Tokushima University, Construction Engineering Department
1996~2000 Moo Architect Workshop
2000 Established Shogo Aratani Architect & Associates
2005~2009 Instructor at Osaka City University
2010~ Instructor at Kyoto Seika University
2015~ Instructor at Kobe Shoin Women's University

Takanori Ineyama Architects

p.108

Takanori Ineyama

1982 Born in Yamanashi
2005 Graduated from Tokyo University of Science
2006~2009 Warked at Architectural design office in Yamanashi
2009~2014 Warked at ON design
2014 Established Takanori Ineyama Architects
2015~2018 Graduate assistant at Tokyo University of Science

Takeru Shoji Architects

p.170, 246 / p.012

Takeru Shoji / First Class Architect, Member of The Japan Institute Of Architects

1975 Born in Akita, Japan
1999 Graduated from the University of Niigata, Bachelor of Engineering in Architecture
2002~2003 Berlage Institute, Rotterdam, The Netherlands
2004 Co-established siki architects studio
2008 Established Takeru Shoji Architects. Co., Ltd.
2012~Present Part-time professor at Nagaoka institute of design

Takeshi Hosaka Architects p.020, 146

Takeshi Hosaka

1975 Born in Japan
1999 Bachelor Degree of Architecture from Yokohama National University
1999 Established Architecture Speed Studio/ Principal
2001 Master Degree of Architecture from Yokohama National University
1999–2004 Speed Studio / Principal
2004~ Established Takeshi Hosaka Architects/ Principal
3,April ~ 3, may ,2013 Exhibition "Ku u so u" at Czech Ceske Budejovice House of Art Museum in Czech Republic
7~10, April 2013 Lecturer and work shop at Faculty of Art and ArchitectureTechnical University Liberec in Czech Republic
2015~ Associate Professor at Waseda University

TAKiBI p.266

Akihiko Tochinai / *Left side*

1980 born in Kanagawa, japan
2004 graduated from Shibaura Institute of Technology
2006 completed master's degree at the Shibaura Institute of Technology
2006~2017 KAZUHIKO NAMBA+KAI WORKSHOP
2017 established TAKiBI

Kiyo Sato / *Right side*

1980 born in Fukuoka, japan
2004 graduated from Kyushu Institute of Design
2006 completed master's degree at the Kyushu University
2006~2010 SHOTENKENCHIKU-SHA Publishing co.,ltd
2017 established TAKiBI

Tato Architects p.034

Yo Shimada founded Tato Architects in 1997, after he graduated from Kyoto City University of Art. As a student, he completed a number of designs and build architecture projects at the school. Based in his home town of Kobe, Tato Architects undertakes projects throughout various locations in Japan. The projects are mainly private houses; however, in recent years the office has taken on assignments for interior and installation work as well. The ideology of the practice is to formulate positive outcomes through understanding the subtle conditions and constraints of everyday life, which exist within notions of place, culture and the ongoing history of architecture. These ideas have been further discussed in publications such as, 7iP #04 YO SHIMADA(7inchproject, 2012)and Everyday Design Everyday : Yo Shimada (Contemporary Architect's Concept Series 22,2016)

The name Tato (タト)is derived from the decomposition of the kanji character 外(outside), which can be read in multiple ways. Tato Architects is interested in the process of continuous exploration, in order to achieve a kind of ambiguity in its architecture; an architecture that renews itself through the perspective of people's everyday experiences. Furthermore, the practice aims to create an architecture that quietly alters people's consciousness.

Tom Robertson Architects p.106

Tom Robertson Architects are a diverse team, driven by a shared love of exceptional architecture. Tom Robertson began his career with Fender Katsalidis Architects working on major local and international projects. Motivated by his belief in personal, flexible service, he established his own practice in 2013. Design is a process of collaboration. Tom Robertson Architects are dedicated to an approach that is accessible and consultative, helping clients to refine their ideas and articulate what they love. True collaboration is grounded in trust, open communication and enthusiasm for the project. Tom Robertson Architects are not defined by one aesthetic or style, but by an approach that is sympathetic to context and environment. Every project is unique yet founded in the central belief that good architecture brings joy. Working with a restrained palette of natural materials, Tom Robertson Architects find elegance in simplicity. Employing refined design that is rooted in pragmatism and driven by beauty, the practice is focused on creating enduring, contemporary architecture.

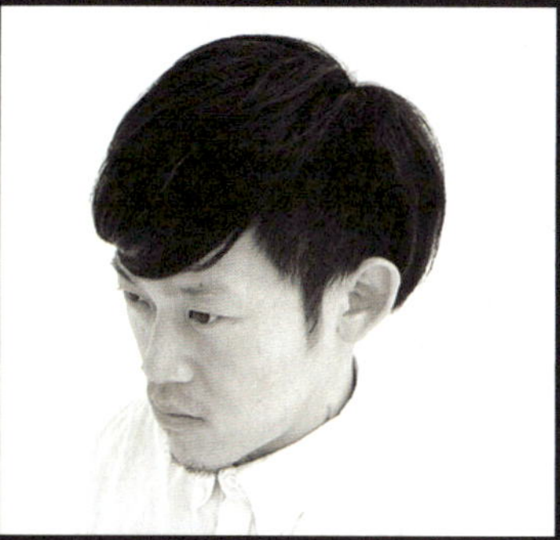

Tsukagoshi Miyashita Sekkei + Keitarchi p.232

Tomoyuki TSUKAGOSHI / *Up side*

1983 Born in Tokyo
2006 Graduated of Department of Architecture,Tokyo Institute of Technology
2006~ Studied at Delft University of Technology
2009 Completed a master's degree, Graduate School of Engineering,Tokyo Institute of Technology
2009~ Worked at Yasutaka Yoshimura architects
2013 Established Llc.2K54DESIGN
2015 Changed the name to TSUKAGOSHI MIYASHITA SEKKEI
Lecturer Nippon Institute of Technology
Town planning advisor in Hachioji-city
2019~ Lecturer Utsunomiya University

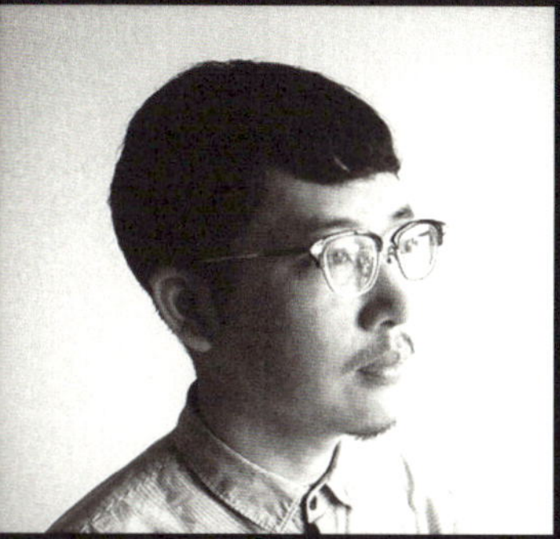

Jumpei MIYASHITA / *Down side*

1983 Born in Kanagawa
2007 Graduated of Department of Architecture,Tokyo Institute of Technology
2009 Completed a master's degree, Graduate School of Engineering,Tokyo Institute of Technology
2009~ Worked at MIKAN
2014 Inaugurated as the Representative Person

Urban Agency p.070 / p.260

URBAN AGENCY is an award-winning architecture firm with a broad international profile. The office is based in Copenhagen, Dublin and Lyon with projects ongoing throughout Europe, Asia and North Africa. It is led by 4 partners – Maxime Laroussi, Andrew Griffin, Henning Stüben and Heechan Park. URBAN AGENCY takes pride in devising innovative solutions to address contemporary challenges in cities and the built environment generally. In designing pragmatic, sensitive and innovative solutions, we create new possibilities for a better everyday life. We design flexible, dynamic and powerful projects that can absorb complexity and change, while maintaining and building-up on existing qualities. URBAN AGENCY's value add is through design. Values add for the betterment of society, as well as the individual client. We have a proven track record of adding monetary value to new and existing buildings. Through careful analysis, programming and massing, we design innovative mixed-uses providing additional square metres for productive and content end-users and stakeholders. URBAN AGENCY believes in creating spaces where individuals can interact – aesthetically inviting spaces of social encounter. We develop comfortable and ecologically responsible solutions that create new value and identity and are economically profitable.

YH2 Architecture p.110, 164

YH2 is an architectural design firm founded in 1994 by architects Marie-Claude Hamelin and Loukas Yiacouvakis. yh2 is a Montreal architecture firm founded in 1994 by architects Marie-Claude Hamelin and Loukas Yiacouvakis. For yh2, architecture is the art of place-making. It is a creative and transformative tool, a way to reinvent reality. All aspects of a project are carefully studied: integration within a given context, conceptual design, working drawings, architectural details, interior design and object design. Concepts are developed in such a way that each of the elements must play a part in the overall composition. Nothing is superfluous. yh2 deliberately chose to concentrate on fewer projects. Headed by its two founding partners, the team dedicates its full attention to these projects, which have received numerous awards throughout the years.

Yoshichika Takagi + associates p.212 / p.034, 076 / p.066, 086

1975 Born in Sapporo Hokkaido, Japan
1998 Graduated Hokkai Gakuen University
2000-2001 Architectural Association School of Architecture
2005-2012 Sekkei-sha inc.
2012 Yoshichika Takagi + associates
Part-time teacher in Hokkai Gakuen University
AR Awards for Emerging Architecture 2010 RUNNER-UP